Advance Praise for *Steadfast*

"In *Steadfast*, James Reyes-Picknell has presented a compelling case for making asset management and reliability a strategic prerogative. The book describes how a culture of excellence in asset management can lead to increased productivity, safer operations, and boost workforce morale as well. It also outlines the importance of leadership in establishing that culture."

—ROY SLACK, Past President of Canadian Institute of Mining

"James Reyes-Picknell has made it his vocation over 40 plus years to successfully help enterprises of all types and sizes, achieve equipment reliability and financial improvement through the implementation of proven maintenance practices and processes.

Steadfast draws on James's lifetime of being a reliability improvement practitioner and teacher by delivering one of the most comprehensive, authoritative books on the topic of how equipment reliability improvement remains a critical contributor to enterprise financial success.

This is a must-read for managers of all levels and functions in Enterprises that have a quest for continuous improvement."

—GINO PALARCHIO, Retired General Manager Manufacturing Services Arcelor Mittal Dofasco

"*Steadfast* captures what operational leaders live every day: reliability is a leadership choice. James makes the connection between disciplined maintenance, empowered people, and predictable performance crystal clear. This is a practical, credible guide for anyone leading complex operations and I believe a lot of organizations can benefit from this work."

—MARIANA PINHEIRO HARVEY, P.Eng., General Manager, Macassa Mine

"*Steadfast: The Reliability-Driven Organization* shows how to turn reliability into an organizational capability through disciplined execution and deliberate skill development. It connects reliability strategy to workforce competence in a way many organizations miss, and is especially relevant for LATAM leaders who must build capability through education, certification, and standardization while operating under resource volatility."

—**GERARDO TRUJILLO**, Consejero de Noria Corporation, Fundador del Congreso de Mantenimiento & Confiabilidad, Noria LatAm

STEADFAST

The Reliability-Driven Organization

James V. Reyes-Picknell

INDUSTRIAL PRESS, INC.

Industrial Press, Inc.

1 Chestnut Street
South Norwalk, Connecticut 06854
Phone: 203-956-5593
Toll-Free in USA: 888-528-7852
Email: info@industrialpress.com

Author: James V. Reyes-Picknell
Title: Steadfast: The Reliability-Driven Organization
Library of Congress Control Number: 2026939324

ISBN (print): 978-0-8311-3704-5
ISBN (ePUB): 978-0-8311-9686-8
ISBN (eMOBI): 978-0-8311-9687-5
ISBN (ePDF): 978-0-8311-9685-1

Publisher/Editorial Director: Judy Bass
Compositor: Patricia Wallenburg, TypeWriting
Proofreader: David Johnstone
Copyeditor: Maki Wiering
Cover Designer: Jeff Weeks

books.industrialpress.com
ebooks.industrialpress.com
1 2 3 4 5 6 7 8 9 10

To my clients—you have taught me so much.

And

To my wife, Aileen—you inspire me to achieve so much more.

I love you.

CONTENTS

FOREWORD

I first met Jim Reyes-Picknell in Toronto in the late 1990s. He was delivering reliability training for our Canadian gold mines. I was VP of Operations, learning from him about how machinery and plant equipment fails, what we can do to avoid those failures and avoid the unwanted consequences they bring. Our mines latched onto the ideas and started to apply them—it helped our production. His advice about maintenance at a couple of the mines was a bit hard to swallow, but it paid off to follow it.

A few years later, I was working as VP Operations at a large nickel operation in Indonesia. We were struggling with production and equipment problems and well behind on production targets. With Jim's help, we increased to nameplate and beyond. achieving nearly 50% increase in production over the next couple of years, and all of it through a focus on maintenance and reliability.

Several years later, I was being appointed as Managing Director of a large diamond mining operation in southern Africa. I called Jim with a now familiar request—can you help us? Things are a mess there. He helped again. I've been impressed by the results he gets with his focus on maintenance, a key activity that is often poorly understood, overlooked, and under-appreciated.

During those efforts, Jim didn't shy away from telling me what he felt I needed to know. Even if he knew he was going counter to what I was thinking or about to do, he spoke up. In Africa he helped me avoid a decision about manpower that could potentially have had disastrous effects on our output. He showed me an alternative approach that would help us improve as we gradually reduced manpower through attrition. He's a straight shooter and as a consultant who truly has his customer's best interests at heart.

Steadfast describes both the business benefits of doing the right things the right way, and an overview of how they are achieved. He focuses on efficiency in maintenance and effectiveness of the work being performed. The result, as I've seen, is higher reliability that translates in increases in production, enhancements to safety, lowering of operating risks, and better overall financial performance.

This book is aimed at the senior executive who needs high performance from his or her physical assets and is struggling to achieve it. It's written in an easy to read style. It should be easy for financial, operational, and other non-technical executives to understand.

There are a lot of gems in this book. In these days of increased public pressure to operate safely and clean, regulatory oversight, shareholder pressure, increased public awareness of industrial risks, holding senior executives accountable for what goes wrong in their companies, and bad news that travels ultra- fast, this is a must read.

James K. Gowans

PREFACE

Maintenance has come a long way—and it's still evolving. We started with the simplest idea of all: Fix it when it breaks. Then came the Industrial Revolution, and we learned that preventing breakdowns was smarter.

The third phase, sparked by the aircraft industry, brought strategy into the mix—balancing preventive, predictive, and detective methods with an understanding of consequences and risk.

Now we're in the fourth phase—evidence-based, data-enabled, and powered by technology. But let's be honest: it's a bumpy ride. Too often, the basics are forgotten in the rush toward "smart."

Many organizations are still stuck in the old ways. A few have mastered the third phase. Too many are trying to jump straight into the fourth and getting lost along the way. The result? Confusion, frustration, and wasted potential. It's time to bring some practical sense back into the picture.

Existing books on maintenance and reliability focus on the technical aspects of keeping things running, methods, and managing a department. While implied, they don't explicitly link to and explain how to maximize business value.

In practice, achieving sustained high performance in physical assets isn't just a technical endeavour. Senior leadership levels need to take an active interest and, critically, need to be engaged in the effort. Serving the owners' interests requires an alignment among the various organizational silos that is achieved only through executive leadership.

Technologically, a lot has changed. Machine learning, artificial intelligence, and data analytics have transformed how we think about asset management. Unfortunately, many are still struggling to get real value out of these tools. The technology isn't the problem—it's how we use it. We've become so reliant on systems designed by people who know the tech but not the work, that we've lost some of our own ability to manage effectively.

When it comes to leadership, there needs to be more emphasis on managing and on the strategic business value that maintenance, reliability, and asset management deliver. That is seldom fully understood at the executive level. My experience over the last 30 years as a consultant is that without that understanding, efforts to improve in this area will deliver mediocre results at best. That's simply not good enough in today's business environment.

I've written other books that show the business value of reliability, but here's the truth—the message often gets drowned out by short-term thinking. In many companies, the focus is still on quarterly results, not long-term performance. And that's where we lose enormous potential. A book alone can't fix that, but it can light the path forward—for executives, managers, and educators who are ready to see maintenance as a true strategic advantage.

This isn't just another technical or how-to book on maintenance and reliability; it's guidance for executives who truly want to achieve the benefits of sustained high asset performance with no surprises because some detail gets lost in the cracks between organizational silos.

It's organized around three key audiences:

1. Maintainers, who need a strong practical base
2. Managers, who must lead teams and drive improvement
3. Executives, who need to see maintenance as a strategic business lever

This book is about turning maintenance into strategic business value.

- Part I builds a shared foundation without a lot of technical detail and how-to.
- Part II dives into management and leadership.
- Part III explores maintenance as a driver of strategy and profit.
- Part IV is about transformation—how to make *Steadfast* performance real in your business.

It's about bridging the gap among technology, people, and purpose. And it's about taking asset performance—and excellence—to the next level.

INTRODUCTION

Rethinking Maintenance

For much of the industrial era, maintenance was viewed as a necessary expense—something to control, contain, and justify. The focus was on fixing what broke, as quickly and cheaply as possible. Even as preventive and predictive methods matured, the mindset remained reactive at its core: maintenance existed to minimize downtime, not to enable performance.

That view no longer fits today's world. Across every sector, asset reliability, energy performance, safety, and sustainability are now strategic issues. Boards and investors ask not only what failed but why, and what that tells us about the resilience of our systems. Maintenance, once buried in the plant basement or service truck, now sits at the intersection of technology, data, and enterprise value.

This book, *Steadfast*, is about that transformation. It explores how organizations can move from maintaining assets to managing their performance, from firefighting to foresight, and from cost-center thinking to a culture of reliability that drives results.

The Four Stages of Maintenance Capability

To understand where organizations stand today, it helps to look at how maintenance capability has evolved.

- **Stage 1—Reactive.** Maintenance is a cost to control. Work is driven by failure; success means restoring operations quickly.

- **Stage 2—Planned.** Maintenance is managed through schedules and preventive routines. Control improves, but resources are often used inefficiently.
- **Stage 3—Proactive.** Reliability becomes the focus. Data, analysis, and precision maintenance enable the prevention of failures before they occur.
- **Stage 4—Strategic.** Maintenance becomes integral to business performance. Decisions about assets are aligned with enterprise goals, risk management, and sustainability.

Steadfast focuses on this fourth stage—the strategic capability that connects technology, process, and people. It's where digital tools and analytics are not just bolted on but woven into how the organization learns, plans, and performs.

The new frontier is not simply predictive maintenance—it's precision performance: doing the right work, the right way, at the right time, for the right reasons.

Why Now

The drivers are clear and urgent:

- Aging infrastructure demands smarter investment decisions.
- Digital transformation promises efficiency but often overwhelms without strategy.
- Energy and environmental goals require optimized asset life cycles.
- Workforce transitions are reshaping how we attract, train, and retain technical talent.

In this environment, traditional maintenance philosophies are no longer enough. Leaders need a roadmap that connects the shop floor behavior to the boardroom goals—linking reliability practices to business outcomes, sustainability targets, and corporate purpose.

Steadfast offers that roadmap. It translates decades of proven reliability principles into a framework for the digital age, balancing technology adoption with operational discipline and human judgment.

What You'll Find in This Book

Steadfast is structured to guide readers from understanding the challenge to building capability and sustaining results.

- Part I explores the evolution of maintenance and reliability—how organizations progress from reactive control to strategic capability.
- Part II examines the enabling tools, technologies, and methodologies that make precision and predictability possible.
- Part III brings these elements together, showing how strategy, leadership, and culture turn maintenance into a source of sustainable competitive advantage.
- Part IV provides a practical roadmap for putting *Steadfast* into action—outlining the steps to assess current capability, plan transformation, and embed reliability thinking into the fabric of the organization.

Throughout the book, you'll find insights drawn from practice, case experience, and cross-industry evidence that demonstrate how reliability principles deliver measurable value. The goal is not to prescribe a single model but to equip you with the structure and insight to build your own approach—one that fits your organization's purpose, assets, and people.

In some parts of the world, such as Latin America, Eastern Europe, the Middle East, Africa, Central and Southeast Asia, the availability of parts, materials, rapid deliveries, and even skilled labor cannot be taken for granted. For readers who may be in regions where input resources may be limited or volatile, keep in mind that *Steadfast* is not a technology-first journey. It is a discipline-first journey. In settings where budgets, skills availability, parts lead times, and leadership continuity vary widely, progress depends on sequencing: *stabilize, standardize, then digitize.* In these contexts, the fastest gains often come from foundational practices—planning and scheduling, precision basics, lubrication and contamination control, storeroom discipline, and cross-functional governance—before advanced analytics. This book highlights both the destination and practical pathways for different starting points.

A Cross-Functional Imperative

No single department can deliver *Steadfast* performance alone. Success depends on collaboration across operations, engineering, information technology (IT), finance, and sustainability. When reliability becomes a shared value, the organization can align around asset performance as a unifying measure—one that bridges production efficiency, safety, and environmental stewardship.

For executives, this means reliability must be built into strategy and governance, not left solely to maintenance teams. Their choices about capital, metrics, and culture determine whether maintenance is seen as a cost or a capability. *Steadfast* reframes it as the latter—a strategic lever that drives growth, resilience, and long-term value.

True reliability is never driven by one tool or department. It emerges when leadership, structure, process, people, data, and technology work in alignment. This is the *Steadfast* Alignment Stack—a simple picture of how maintenance becomes strategic.

Steadfast	Purpose and business strategy
Alignment Stack	Governance and organiation structure
	Processes and standards
	People and culture
	Data and information
	Technology and tools
	Performance outcomes

FIGURE I-1 The *Steadfast* Alignment Stack—the System Required for Reliability Excellence

Throughout this book, we return to these layers again and again, because the real power of *Steadfast* lies in aligning them—not in optimizing any single one in isolation.

Looking Ahead

The journey to *Steadfast* performance is not a single initiative but an organizational evolution. It begins by changing how we think—seeing maintenance not as the end of failure but as the foundation of performance. It continues through disciplined steps that integrate people, process, and technology toward a shared purpose: reliable, efficient, and sustainable operations.

This book invites you to view maintenance in that light—not as a technical afterthought, but as a strategic advantage. *Steadfast* is where reliability meets leadership, and where every decision about assets becomes a decision about the enterprise's future.

PART I

FOUNDATIONS OF MAINTENANCE EXCELLENCE

A physical asset that is well designed, well operated, and well maintained adds value related to its performance in the business. Do any of those three poorly, and the value your physical assets deliver drops fast.

Maintenance is about one thing: sustaining the performance of physical assets. Engineering designs them; maintenance keeps them performing as designed. When design flaws show up, maintenance often spots them first—and sometimes even finds the fix. If the asset fails due to factors outside of its control, such as errors in operating the equipment, then it is quite possible that even the best maintenance cannot change the situation without the collaboration of operators and others.

As an operational function in a business that relies on its physical assets, the maintenance department is largely focused on doing. It does proactive maintenance to avoid or reduce the consequences of failures that are otherwise inevitable. It does reactive maintenance, fixing equipment and systems that have been allowed to fail. Doing precision work with skill requires education and training, often beyond that taught in trade and engineering schools. It executes its work with established plans, on schedule, and with precision. Part of that execution is the troubleshooting of failures and identification of just what has gone wrong, and the analysis of failures after the fact with an eye to avoiding them in future. All of that requires up-to-date and accurate information about the equipment being maintained and its parts, as well as knowledge of its importance to the business. That involves a coordinated effort on the part of maintainers, operators, planners, schedulers, information management, engineering, and spares management. That coordination is what separates good maintenance organizations from great ones—they work as partners, not as isolated crews.

This part of the book is primarily for maintainers. It provides a common baseline understanding of the maintenance and reliability function.

CHAPTER 1

THE CHANGING ROLE OF MAINTENANCE

Phase 1: Break—Then Fix

As soon as mankind learned to use tools, we've had to learn to repair or replace them when they break. Repair is the most basic form of maintenance—waiting until your equipment breaks, then restoring it back to that "good as new" or "good enough to be used again" condition.

Early man didn't understand thermodynamics and deterioration into chaos as we do today. We know that if we don't put energy into our tools and systems, they will naturally deteriorate. We have learned to improve designs and look after things so they continue to function as we intend.

Design Out Weakness

By paying attention when things break, we learned about their weaknesses. Creative ideas were used to strengthen those weak points, make them less prone to breaking, and add capability that wasn't in the previous iteration of the design. Wheels got more precisely round, they got lighter, they were less prone to breakage, tires were added to protect the wheel, to help make the ride more stable, and materials got lighter and stronger, as we can see in racing bicycles today. Fire enabled both cooking and seeing more after dark. When enclosed, it cooked hotter and better. For light, it became portable in the form of torches; its light was focused using transparent enclosures and lenses. In time, the fire gave way to electric filaments, and today's light-emitting diodes (LEDs).

Designs evolved and so has our perspective on keeping those designs functioning.

Early Preventive Maintenance

We learned that lubricants like animal fat could make sliding and rotating contacts smoother, easier to move, and less prone to wearing out. We learned that precision in manufacture and assembly minimized wobble and wear. We learned that by evenly distributing weight, we lower wear rates and could use less materials at higher levels of stress. By keeping those lubricants in place, sustaining the precision of fits, and reducing wear, our equipment would last longer. A sword or an axe kept sharp worked better than one that was dull, so instead of throwing out the old dull ones, we learned to sharpen them to keep them functioning. We also learned that if we did that before the blade was dull, it was easier to sharpen, and it functioned more consistently.

The concept of preventing failures came along quite naturally. Some failures, particularly those associated with mechanical wear, can be prevented, and in doing so, we lessen their impact on our use of those assets, and increased the benefit they provided.

The Need for Prediction Soon Followed

If the mechanism of failure can be detected early enough, we can even time that preventive work to minimize impact even further. Preventive action avoids the failure, but it can be costly because those failures don't always happen at the same time. We need to act early and sometimes "waste" whatever useful life might be left in the asset. By monitoring its condition (e.g., how sharp the blade is), we can increase its utility and sharpen it only when it really needs it. Although it wasn't called by its modern name, that inspection and monitoring was an early form of predictive or condition-based maintenance.

Don't Forget the Safety Devices

Early systems rarely came equipped with safety devices. As equipment got bigger and had higher energy density, we realized that some failures can result in disastrous consequences like injury or death. Making sure those devices were in good working order was often overlooked, occasionally with disastrous results. It wasn't until reliability-centered maintenance was introduced in 1979 that making sure those devices could work when needed was really addressed.

Industrial and Maintenance Evolutionary Stages

More recently, we have the concepts of four stages in the Industrial Revolution. The first was mechanization with steam or water-power-driven machines. The second was the introduction of mass production and electric-powered machines. The third entails the introduction of electronics and information technologies to help in controls and management. The fourth takes advantage of cyber-physical systems, hybrids involving advanced Internet of Things and machine learning.

From the foregoing discussion, it's clear that maintenance has undergone a similar evolution: break—then fix, prevent, reliability-centered (heavy on predictive), and now predictive-enhanced with data-enabled (often called "data-driven") decision-making tools.

Phase 2: Preventive

As new technologies emerged, maintenance management had to change—fast. Initially we had craftspersons who often operated their machines and then fixed them when they broke. They learned to prevent failures that were preventable, to make their tasks easier. As production levels and complexity of machines and factories grew, we introduced specialization in both maintaining and operating. The concept of preventing failures spread, and in larger organizations we had preventive maintenance teams or technicians as well as those who repaired what broke. Proactive (preventive) schedules were not complex, nor did the work apply to everything. Management was pretty straightforward and often carried out by technicians who demonstrated good supervisory or management skills.

Phase 3: Predictive

Predictive maintenance was once very simple, high-touch and low-tech. Technicians could hold a screwdriver to their ear and place the tip on a bearing housing to hear the condition of a bearing. They could put their hands on equipment to see if it was running too hot. They could listen for odd sounds indicating that something was amiss. Oil could be sampled and visually checked for contaminants or excessive water content. Oil analysis got somewhat more sophisticated using spectrographic methods to detect wear particles in oils lubricating critical equipment. As electronic technologies advanced, so too did the ability to monitor equipment conditions with vibrations, thermal imaging, and ultrasound. The provision of predictive maintenance tools, methods, and services was becoming an industry of its own. Larger organizations started to have their own predictive maintenance technicians.

Phase 3 Plus: Reliability-Centered Maintenance

In the late 1970s, the term "reliability-centered maintenance" (RCM) was coined in a study[1] done on aircraft failures. It added considerably to the science behind failures, how they manifested and what could be done about them. It recognized and quantified failures as being either age-related (good candidate for preventive maintenance), randomly occurring (candidates for predictive maintenance), or prone to infant mortality (subject to quality controls and risky to disturb). It also introduced the concept of "failure finding" for devices that were usually dormant, failed randomly, and were needed to act when another failure or condition required control (e.g., safety devices).

The military in the United States and United Kingdom led the charge. They quickly saw how RCM could boost readiness—and it worked. I used a military version of RCM in large defense projects in the 1980s. John Moubray[2] and Anthony M. Smith[3] were the first to commercialize it in the early 1990s. By the late 1990s, awareness had grown, and its use had spread to most other industries, including nuclear, food, pharmaceutical, utilities, manufacturing, pulp and paper, mining, and elsewhere. Today it is widely regarded as the gold standard in how to identify the most appropriate failure management strategies.

With RCM, the focus in maintenance management was expanded beyond repairs and prevention. It now embraced predictive work, and we had the concept of proactive maintenance programs. Reliability was now a focus. In those early days, applications of electronic technology in condition monitoring and management information systems were expanding rapidly. RCM helped us focus those monitoring applications where they would do the best. Management of the maintenance function started becoming more sophisticated. It grew from oversight of technicians who did repairs, to include planners, technologists and engineers in multifunctional departments.

Phase 4: Asset Management

The concept of asset management then came along in the early 2000s. It was a result of the need for more consistency in how entire portfolios of assets such as electric, gas, and water utilities were managed. The additional finance and forecasting became essential to the management of these assets. Asset management is a broad field dealing with the entire life cycle of assets—it spans maintenance, reliability, engineering, operations, and finance. Over the past 20 years, asset management has grown to be accepted in many organizations for its strategic focus on finances blended with asset life-cycle activities. That role tends to be centered in the head offices of larger organizations and out of the hands of the operating sites who still manage maintenance directly. Asset management is even mandated by law in some jurisdictions due to its value in managing public funds wisely.

Computers and the Rise of Evidence-Driven Decisions

Computerization and information technology expanded rapidly in the 1980s and onward. Greater predictive capabilities arose and were put into practice, and technology became pervasive. In the 2000s, the concept of making better "evidence-based decisions" arose from the need to improve asset performance and availability. The vast stores of data about equipment and equipment maintenance were seen as valuable sources for insight into making better decisions. Sadly, much of the data that had been (and continues to be) gathered is of questionable quality. To make good decisions, you need clean data. Unfortunately, filtering out irrelevant entries is tedious work. Few do it—and those who try often struggle.

Industry 4.0: Where We Are Now

Technology now enables sensors to be installed anywhere and interconnected via the internet or other networks. Those devices often include some computational ability at the sensor so networks aren't flooded with "normal" signals when we really want only "alerts" when abnormal situations arise. Many of these devices are for condition monitoring. Data analytics expanded, and the dream of having the machines monitor machines seemed real but often fell short of expectations. With the expansion of machine learning and artificial intelligence, we now have rapid growth in capabilities to use data, learn, and apply what's known to new problems. Machines, programmed with rules and looking for patterns in data, are now very good at diagnostics. This is still all pretty new, and there is little education and guidance on how to manage and leverage it. The technology folks are charging ahead, while most of us end up confused, struggling to make sense of it all.

It's hard to embrace what we don't fully understand—and that's why so many organizations remain stuck in phase 1 (reactive), 2 (preventive), and 3 (predictive). We are not yet fully leveraging RCM (phase 3 plus) nor the technology-enabled capabilities that characterize phase 4.

Today's Challenges

I see this all the time: managing it all has become quite challenging. "Old school" technician-based management thinking and experienced trades without specialized education are not equipped to manage today's complexity. Even engineers who have not kept up with all of that development find themselves challenged to manage it effectively. I also meet managers with only business school education. They are usually lost in the world of maintenance and simply leave it to the technicians—the uninformed are leading those who are often less informed than they need to be.

We need to understand the reliability-centered concepts and then apply them. Outside of maintenance, and sometimes inside, those concepts are not understood. In today's fast-paced business environment, there is also continual short-term pressure on operations: Produce more and faster. Quarterly shareholder value considerations drive short-term thinking. That undermines many efforts at continual improvement. Like the classic story of the tortoise and the hare, production runs like the hare, breaks down, waits for the fixes, and ultimately slows output. The concept of slowing down to go faster,[4] like the tortoise, putting less stress on the assets, having fewer breakdowns and increasing overall output, continues to elude many.

Aside from obtaining needed parts in time, one of the biggest battles maintainers have is with their operations counterparts, short-term quotas, and "Everything is urgent" mindset. They simply don't want proactive maintenance to be done if it requires machine downtime. The work gets deferred, and lengthier breakdowns follow.

Technology-related challenges have also emerged. Myriads of technologies are available. Managing those, particularly in an increasingly connected environment, increases risks of sabotage by cyber criminals. Information technology management is more aware of the risks and tends to put the brakes on our use of many technologies that can actually help us. Information technology seldom makes life simpler with industrial operating and information management systems. Those systems are not easy to use and fall into disuse. The system becomes the focus, not the process nor the goal.

Technology has driven specialization. Decisions often involve multiple disciplines, and sometimes we don't appreciate the need for all the involvement. Very little in our world of maintenance is entirely within our control, yet we are held accountable for results despite our siloed organizations that discourage collaboration.

The heavy reliance on computerized management systems has switched our focus from "processes" and "results" to "details" on screens. We are looking at the leaves, forgetting we are in a forest and there's another important forest nearby that we must interact with. Interestingly, all this technology connects us when it comes to information and data, and puts more space between us when it comes to real communication.

Modern management systems are complex and highly integrated when it comes to sharing data. They are expensive and difficult to implement. Technology projects are often delayed or run out of budget. Training is often sacrificed on the altars of schedule and budget. Understanding of the processes being automated and how to use the technology suffers. Users muddle with their systems until they find something that works for them, whether or not it follows intended business processes. Then they teach whatever they have found, often bad habits, to others.

I've watched too many organizations race to adopt technology before mastering the basics. We have capable tools, talented people, and more data than ever.

Yet, we still struggle to use them well. We're tech-enabled but data-drowned—and too often blinded by urgency. The challenge ahead isn't more technology; it's learning to manage what we already have, with purpose.

CHAPTER 2

PRINCIPLES OF MAINTENANCE EXCELLENCE

Excellence is a journey—a steady climb toward mastery. In maintenance, that journey starts with intent: knowing you want to improve, learning the basics, and practicing until you get it right—every time. In maintenance, we are taking actions that counter the very natural tendency of all things to degrade.

The most successful maintenance programs are designed with consideration of the failure mechanisms, triggers, consequences, how to most successfully deal with those, and being efficient in executing that maintenance.

Degradation and Mitigation

Engineers study thermodynamics and we learn its second law: entropy (a measure of chaos, randomness, or disorder) of a closed system can never decrease over time. It either increases (becomes more disorderly) or remains the same. Entropy never sleeps—you can only fight it by adding energy. Our built environments—buildings, infrastructure, processing and manufacturing plants, equipment, physical assets of all kinds—are all subject to that second law. We build them to perform a function, and then they naturally move toward chaos. We need to act to arrest that movement. To do that, we add energy, in the form of maintenance.

The kind of maintenance we need depends on what's happening inside the asset. Is it erosion, corrosion, fatigue, or simple wear and tear? Maybe it's a random event that starts a chain reaction. Each one needs a different response. To deal with erosion, we make surfaces more resistant to wear. To deal with corrosion, we change materials to those that don't corrode, or we coat them to alter the chemistry. With wear and tear, we build assets with greater resistance, or we restore those surfaces that are degrading. Random events cannot be prevented. They happen at any time, some with immediately disastrous effect. Others trigger degradation, and that will happen over time, once initiated. If we understand the nature of the degradation, we can take action.

The right maintenance depends on what is actually happening in the asset, how it is triggered, and how rapidly degradation occurs. That's the essence of good maintenance: understanding what's happening before deciding what to do.

When selecting the best maintenance approach, we also need to understand the consequences, or impact of the failure, and whether or not it is worth dealing with. A failure that has a big impact on the function being performed, such as loss of production or inability to deliver a service, is vital to the business. A failure that results in the automatic start of a backup, with no loss to production, can be tolerated. If a failure can damage the plant, harm someone, or spill something toxic—it's worth preventing or at least taking steps to mitigate its consequences. A failure to a safety device, such as an electrical breaker, can be tolerated so long as it isn't needed, but if it is, testing to make sure it can work is needed to avoid the consequences.

We've looked at how assets degrade and how we counteract that. Now let's talk about what we actually get in return: availability.

Availability

Whenever we build an asset, we expect it to do something for us—its function. Usually, it has more than one function. For instance, a pump is there to provide a flow of a liquid. To do that, it must also contain the liquid, another function, and have sufficient driving power to achieve the flow and pressure rise (another function). If that driving power is a result of electrical energy conversion in a motor, it must also prevent electrocution of those nearby (yet another function).

We want those functions working all the time, or at least as much as possible given that entropy is always working against us. That portion of time when we can use the asset is known as "available" time, whether it is being used or not. The measure of available time as a portion of total time is known as "availability." We want availability to be maximized.

We do that by reducing the number of times the item breaks down (reliability) and also by reducing the time it is down for repair (maintainability).

Reliability

An item that doesn't breakdown often is considered to be "reliable." We want high reliability, usually measured by the time between breakdowns (i.e., hours between failures), or its inverse, how often it breaks down (i.e., failures per hour). Reliability results from how well the item is designed for the function it must perform. Simplicity of design, materials used, how well mechanical stresses are distributed, the selection of parts, and so on are all factors. Reliability is designed in and cannot be changed without changing design of the item.

An item that breaks down often is said to be "unreliable."

In service, achievement of reliability is influenced by the correct application of the right maintenance. For example, if you have sliding surfaces, they must be lubricated by the right lubricants and in the right quantities. The sliding surface is a design feature, it has friction, and lubrication is needed to reduce friction and wear. Lubricants are consumed in operation, so keeping it lubricated is a function of how well it is maintained.

If we fail to maintain those necessary operating conditions, we cause the item to be less reliable. In many operational environments, it is not the design that is unreliable, it is our failure to maintain properly that leads to unreliable operations. Execution is where we make or lose reliability—and that's entirely within our control. The application of the right maintenance, at the right time, and in the right way, is also within our control.

Reliability is about how often things fail. Maintenance helps us sustain reliable performance. Our ability to maintain is known as "maintainability." It is about how efficiently we can do those needed maintenance actions like lubricating or fixing when they break. Both drive availability (uptime).

Maintainability

Whenever an item breaks down, it must be repaired. The longer it takes to repair, the longer it will be out of service, unable to perform its function. We want repair times to be short, and those are a function of how "maintainable" the item is. Maintainability is influenced by complexity of design. Ease of disassembly, access to remove parts and replace them, sufficient lifting capacity for heavy parts, clear removal routes to get the broken parts out and the new ones in, enough room for maintainers to work on it, and even the availability of replacement parts all impact on repair time. All of those are influenced by design, availability of repair plans, tools, skilled trades, and parts—and all of those are within our control.

If it takes forever to repair, it's not very maintainable, simple as that. Availability will suffer!

Safety

Physical assets are built to perform their functions without causing harm to people (safety), and limited harm to the environment. Structures are built to meet certain building standards, some of which are codified in laws (e.g., building codes). Installed systems must meet certain design safety standards such as an electrical code or the requirements of pressure vessel regulations. Electrical and mechanical systems convert energy from one form to another: motion, pressure, temperature, and so on. When they fail, they can release energy suddenly and in ways that are harmful. Control systems keep things running smoothly, but when they fail, they can result in damage to equipment and systems, also leading to harmful situations.

There are two aspects of safety that we are concerned with. The first is the designed in, or inherent, safety of our systems. Like any other function that inherent safety must be maintained. The second aspect entails being safe with the practices we employ while operating and maintaining our systems.

Safety provisions are designed in features that safeguard us and the equipment itself. They include guards on exposed rotating components, electrical breakers, pressure relief (safety) valves, overspeed trip devices, barriers, warnings, alarms, trips, electrical insulation on conductors, thermal insulation on hot piping, grounding provisions in electrical systems, start permissive logic in control systems, sequencing in control systems, and so on. They must all be functioning correctly so they can do their jobs when needed.

When working on equipment, we need to work safely. Broken-down equipment is not in its normally safe state; there are risks that are not normally present. It may have damaged insulation or broken and sharp edges, it may still contain hazardous substances that could not be drained, and so on. The work we are doing may involve the use of force and tools, heat, open flame, electrical measurements, lifting heavy objects, working in confined spaces where gases may be present, explosive environments, working at heights, or under heavy objects. It is imperative that we know of the risks associated with our systems, equipment, the maintenance procedures themselves, and how to deal with them all so we minimize risks to our operators and maintainers.

Designing for high reliability means fewer breakdowns, hence less opportunity for those risks to arise. Designing for maintainability means that there are fewer risks to workers while maintaining or repairing equipment. Designing for safety means that known risks are minimized and that provisions are made for mitigating those risks. Those are all design features.

We must also maintain safely, and that is something we manage day to day, not just at design.

Excellence Is Value

Each of these—reliability, maintainability, and safety—feeds the same goal: keeping assets productive and the business profitable.

Maintenance Execution

Having a good design is a good first step. Having the right maintenance program designed for the asset is the next step. Sadly, that is not always in place. In some cases, it is entirely overlooked. We do have control over those, so it is a matter of choice.

Having our maintenance program executed properly is imperative. There is no point having a well-designed maintenance program that we don't follow. As you know, aircraft have such well-designed maintenance programs that are followed rigidly, complete with checks and sign-offs to be sure that nothing is left out. Imagine if those checks and sign-offs weren't done, there would be no assurance that the aircraft is being maintained. You might feel uncomfortable flying in it if you knew that. Now imagine that the maintenance program weren't even being executed at all. Would you even get into the aircraft?

Not all assets have such serious consequences of failure as aircraft, but consider this: The equipment doesn't know whether it is in an aircraft or your plant. It only responds to the maintenance that it receives, or not. If you want it to run and achieve the reliability it is designed for, you must maintain it and maintain it with the right maintenance and at the right times. Any less will defeat the purpose of having the maintenance program, increase the risks of failure, risks to the equipment itself, production, operations, safety, and the environment.

Airlines don't leave maintenance to chance, and neither should we. The same discipline applies, whether you're maintaining a jet engine, a haul truck, or a process pump.

Maintenance Costs

Accountants see maintenance as a cost. We see it as an investment. The trick is knowing how much to spend, where, and when.

As a rule, expenses are to be minimized, but not if reducing them impacts the ability to produce revenues. Accountants understand maintenance is necessary and accept that some spending is needed. But we often don't fully understand the different types of maintenance, what they cost relative to each other, and how important they are.

Obviously, repairs are important. You can't go without them when something breaks. Because breakdowns are often seen as unavoidable, repair costs are considered "nondiscretionary." It is also useful to know that the cost of repairs is almost always greater than the cost of avoiding them. You must fix things, or you are out of business. However, there is a subtle refinement that many accountants miss. While the repairs are unavoidable, the breakdowns that lead to them can often be avoided, or at least minimized. To do that requires some "discretionary" spending on proactive maintenance and that cost is substantially lower than the cost of repair.

Is Maintenance "Discretionary"?

In the absence of any being proactive, the default maintenance strategy with any physical asset is "run-to-failure" or "break-then-fix." Nature takes care of that; no thought is required. That is a "reactive strategy," and it is very expensive. Breakdowns often cost three or more times the cost of avoiding them.

They result in premature shortening of equipment "life," high repair costs spent on parts, emergency shipping of those parts, inefficient work practices due to a lack of planning, overruns on repair time estimates, and excessive levels of overtime and contracting. In addition to higher costs eroding income, the extended downtime results in revenue loss, higher costs per unit produced, and lower margins.

With more proactive work, there will be fewer breakdowns, less costly repairs, and improved business results. The program becomes less expensive and downtime is reduced, improving revenue capacity. However, it is also possible to do too much of the wrong maintenance and actually make things worse.

Being proactive with some basic tools and rudimentary knowledge of their use pays off. However, it's natural to tend to overdo the wrong type of proactive work. Avoiding that requires knowledge that is often not taught in engineering or trade schools. You will need to spend a bit to educate your maintainers to get the most benefit from being proactive.

"Proactive maintenance" is optional, discretionary, but it is not "nice to have." It is needed if you want anything approaching reliable asset performance. Like exercise and a healthy diet keeping you healthy and out of the hospital, proactive maintenance is administered before the breakdowns with an aim to reducing failure consequences, but not necessarily eliminating all the failures. If you get it right, you avoid the worst breakdowns and their consequences. That minimizes impact on the business. If you get it wrong, you are likely to still have too many avoidable breakdowns, higher costs, and downtime that detracts from revenue generation.

A bit of discretionary spending on the right maintenance can reduce the nondiscretionary portion of spending on repairs, and usually by a factor of three (repairs) to one (proactive) or greater.

The Right Maintenance

If you repair something that isn't broken, you've wasted your effort and introduced the possibility of mistakes that make things worse. If you open things up for inspections, you expose them to dirt and contaminants. If you put the wrong lubricants in or get dirt in with them, you are shortening equipment life. There are many ways to "overmaintain" and lower availability. You want to avoid them.

Operators are often loath to hand equipment over to maintenance for proactive maintenance (PM). They know it often takes maintainers longer to do the work than they first promised. They've also had problems with equipment that was just maintained.

Intrusive inspections are a mainstay in many maintenance programs. They disturb the equipment and often reveal minor defects that maintainers just can't resist fixing, even if they are minor. The equipment had been running well before they looked at it, but they fix whatever they find. That adds to the downtime, cost, and delays it getting back into service. If a mistake is made, or a new part doesn't fit as well as the old worn-in parts, the equipment may break down soon after the maintenance was done. This actually happens a lot. It would have been better to leave the equipment alone. Operators who have experienced that no longer trust maintainers, and understandably so.

They'll tell you that if something breaks, the first place to look is the last place that someone worked on it. They're not wrong either! If the wrong maintenance is done, it is done too often or not often enough, if it is done poorly, or it is skipped when really needed, you get into trouble, and the equipment soon tells you.

The tricky part is to get the right proactive maintenance program implemented. That requires a bit of investment in training (reliability-centered maintenance, RCM), then analysis (takes a bit of time away from the normal job), execution of the program (which usually requires less effort than before), and some careful observation on its effectiveness.

Be aware that the operating conditions and context in which assets work often change and evolve. One electric utility started experiencing more frequent failures after introducing "time-of-use" billing. The changed demand patterns impacted equipment reliability. Maintenance should evolve with the context, something that many miss.

Maintenance the Right Way

In addition to the right program and its execution is the quality with which it is executed. The skills and knowledge of the workforce who performs the maintenance work really matter. If they perform the maintenance the wrong way, they can achieve little or even cause harm. For example:

- If you change oil in an engine and replace it with the wrong oil, you are likely to damage the engine.
- If you replace high-speed rotating components with new ones that are unbalanced, they will vibrate and potentially break when you run them.
- If you clean a rotating component without balancing, it will vibrate badly.
- If you replace a tire with a lower-cost alternative made from inferior compounds, it won't last as long. You will be replacing them more frequently.
- If you bolt parts together with inferior bolts or incorrect tightening, they come loose and damage the equipment.

Doing maintenance work the right way is important to the integrity of the equipment worked on, and hence to longevity, safety, and the other functions of the equipment.

Maintenance supervisors, planners, and trades will focus on these quality aspects first—that's what they control directly. They take pride in their knowledge, skills, and ability to fix things; their creativity in solving problems; and the quality of their workmanship.

Those are key to improvement, but they are but one part of a complex bigger picture.

Continual Improvement

Maintenance is more complex than most people think, and until you live it, you don't know what you don't know. Those who are new to the field quickly learn how much they need to learn and then spend years learning it. Each lesson arises from a new job being performed, and most of those jobs are not repeated often. If they are repairs, there is always something different from the last time, so there is always something new to learn.

If skills have been upgraded, the way you execute a job may be different from how it was the last time. For instance, if you have a new laser alignment tool, the way you align equipment will be different. Each new thing is a lesson that should be documented in the job plan or procedures so it can be repeated the next time—which might be a long time in the future and with someone new.

If the plant has new equipment, it will require new maintenance plans and procedures. Again, these are upgrades to the documentation you have on maintaining the plant.

If you introduce new technology, you may need to upgrade skills in your workforce. For instance, if you introduce oil mist systems to replace oil-filled sumps, maintainers will need to know how to maintain it and stop adding oil to sumps.

Skills don't get better without practice, and some skills in maintenance are not used often. Can your workforce practice, like a rehearsal, before doing a job? Barbers learn to shave other people's beards with a straight razor by practicing on balloons. What do your trades practice on?

When you introduce new management software, or a new policy about workplace harassment, or new benefits packages, do you really teach your people how to use them, or do you just tell them, "It's there" and expect them to read up on it?

New software is one area we often get wrong. We teach people how to use a few screens, but we don't teach them the process that is being automated by it. Consequently, they don't really know what they are doing, nor how their shortcuts and tricks impact the rest of the business. Perhaps they don't care either, but you should, or you'll never get them to do so.

Maintenance excellence isn't a finish line, it's a discipline. Every lesson, every job, every failure teaches us something new. Capture those lessons, refine your methods, and you move closer to mastery, and to truly sustainable delivery of asset uptime.

CHAPTER 3

CORE PRACTICES FOR MAINTAINERS

In later chapters, I'll discuss reliability strategy, maintenance excellence, and asset management. Here, we focus on the foundations: the core practices maintainers need to master to be effective today. They are the foundations for reliability and maintenance excellence that enable us to launch maintenance from its tactical roots to become a truly strategic business function.

Proactive Maintenance Techniques

A poor understanding of proactive maintenance often leads to overmaintaining and misuse of proactive methods.

Maintainers do need to know the basic methods, their differences, and when they may or may not be appropriate to use. We won't go into much depth here. Part II will dive deeper into selecting the most appropriate methods, and you can refer to *Uptime*[1] for more depth on proactive maintenance in general.

Preventive Maintenance

Preventive maintenance reduces failure probabilities by eliminating the failure before it has time (or usage) to occur. These failures arise naturally with age or usage of the asset. Wearing out occurs gradually, and within limits, it can be tolerated. Corrosion, erosion, and fatigue are similar. They all have known limits. We need to monitor the rates at which these phenomena occur, combine that with the known limits, and determine just how long we can operate before we get there. When we reach those limits, we either restore or replace the device before it is "worn out."

Examples include the tightening of drive belts that loosen with usage, wear plates that degrade as material passes over them, replacement of batteries that degrade with use over time, and the replacement of sensors (like smoke detectors) that have limited useful lives. We don't usually monitor their condition; we just replace them. Changing these takes far less time than the repairs that we would have to do if we were to allow them to fail. We benefit from less downtime, lower costs, and less disruption to operations.

Be cautious though: preventive maintenance can be overdone. Replacing components too early adds cost and risk without reducing failure likelihood—a common symptom of overmaintenance. The goal is to balance cost, risk, and reliability, not simply to do more work.

Predictive Maintenance

In predictive maintenance, we take advantage of the fact that many failures (especially in mechanical and thermodynamic systems) give us an advanced warning that they are deteriorating and getting close to the point of failure. Many failures give those sorts of warnings, whether they are failures that arise because of age and usage or randomly.

Detecting Early Signs of Failure

Components that wear out often get loose and make noises or vibrate. They may also get hot due to the extra movement and contact with other components. The noise, vibration, or heat can be detected. Even loose drive belts can be spotted with infrared cameras without removing their guards.

Failures due to erosion and corrosion can often be seen in inspections visually, but if that entails opening up the equipment, it can lead to problems due to mistakes made, dirt ingestion, or even operator errors on start-up after the inspections. Both of those arise because of material properties and exposure to fluids. If those fluids are moving, erosion will disturb the flow, and that may be heard if we listen for it. Like cavitation in a pump, it is very audible and a direct result of flow disturbances.

Many, in fact, most, failures occur due to random causes that trigger the initiation of a failure mechanism. A heavier than normal rain storm might cause some water to get into oil sumps in outdoor machinery. You cannot know when the rain will come, but you can detect its effects if you are watching for them. In that case, oil sampling and analysis could reveal the presence of moisture. It could also detect other contaminants or even wear metal particles from normal operation.

In "random" failures the mechanism of degradation is known, but its timing is not. However, the mechanisms of failure are easily predicted, and we have technologies to watch for them. Infrared, ultrasound, oil analysis, vibration analysis, putting our hands on a bearing housing to feel vibrations and temperature are all methods of condition monitoring. Some are high touch, others are high tech.

Managing the Timing of Predictive Maintenance

With random failures it is important to realize that the timing of when the failure will be found is unknown, so we must check often. When we discover a deteriorated condition, it is telling us we have a problem, the failure has already begun—we'd better act soon. It will only get worse, and usually that happens quite quickly. The time between when we detect the problem and fix it is finite, maybe days or weeks. Our inspection intervals must be shorter than that, or we can miss the warning. We risk not detecting the problems at all if we check too infrequently.

Predictive maintenance doesn't prevent failure; it predicts it. The value lies in acting on those predictions before the problem escalates.

If we find a problem and can fix it early, before the equipment crashes, we can reduce repair costs and likely time the repair work to have minimal impact on operations. But we only get those benefits if we act quickly after we've found the problem.

Detective Maintenance (DM)

While preventive and predictive methods aim to avoid failures and consequences, detective maintenance ensures that protective systems work when needed.

This is less well known than preventive and predictive. It is sometimes called "failure finding testing." It is used where failures occur randomly in devices that are dormant, or inactive during normal system operations.

For example, a high-speed trip on a turbine only works if the turbine overspeeds—a rare event. It protects the turbine from self-destructing and possibly throwing its own parts around the plant. An electrical breaker only trips when it experiences excessive current flows, indicative of some other problem causing an overload.

These devices are there to protect the equipment from damage, to ensure our safety, to protect the environment, or to maintain critical process parameters within a desired operating range. It is important that they can function when needed. Since they are normally dormant, the only way to ensure they can work is to test them.

Those tests are known as "failure finding tests" or "detective maintenance." If the test reveals the device isn't working, we correct the situation and we've lowered the possibility that it will fail" to work when really needed. The testing reduces our risks.

It is possible that these devices fail between tests, so we test frequently enough to lower those risks to levels we feel are tolerable in achieving our business goals. Like the other proactive methods, detective maintenance is only effective when managed deliberately and with clear testing intervals.

Preventive, predictive, and detective methods form the practical basis for reliability improvement and cost optimization discussed later in the book.

When Maintenance Fails—and What We Can Learn

Without the basics and good management of maintenance, we can get into some serious trouble when things go wrong. Here are just a few examples; there are many more.

The Deepwater Horizon disaster (2010) revealed what happens when inspection and testing are neglected: latent failures go undetected until catastrophic results follow. That incident killed 11, injured 17, and released some 4.9 million barrels of crude oil into the Gulf of Mexico with immense environmental harm to wildlife.

The Phillips Petroleum plant in Pasadena, Texas (1989), illustrates the need for proper maintenance practices and procedures. It suffered an explosion killing 23 and injuring 314, as a result of a gas line being opened without being properly isolated.

In 2022, the Morbi pedestrian suspension bridge collapse showed us the importance of using skilled people, using good quality materials and workmanship, doing the job thoroughly, using new materials in critical applications, and applying some basic engineering in planning large-scale work. This bridge collapse in India killed 141 and injured 180. It had just been repaired. Its decking was replaced but not its cables and their anchors or bolts, all of which were badly rusted. The new deck was heavier, and the old structure couldn't take the weight. And in that case, the people who did the work weren't even bridge repair people—they were from a company that manufactured clocks! The wrong people with the wrong skills and knowledge.

These failures remind us that excellence depends not only on methods and tools, but on discipline, competence, and continual learning. Learning from our mistakes is always difficult but well worth it if we don't repeat those same mistakes.

Learning from Failures

Despite our best preventive and predictive efforts, some failures will still occur—either because they were missed or because we accepted the risk. When they occur, we will need to carry out a repair and possibly correct some damage, and we also have an opportunity to learn from the event.

If we observe what events occurred before a failure, what happened during the failure, and the failure mechanisms that occurred, we can gain insight into how we might avoid the failure in the future. That observation and analysis is known as "failure analysis," and it is a part of root cause failure analysis. The failure analysis deals with the mechanisms of failure; the root cause is what set the sequence of events in motion.

The mechanisms of failure give us insight into possible design changes that make our systems more robust and fault tolerant. Dealing with the root cause enables us to stop the chain of events from even starting. In its simplest form, root cause analysis entails the asking of "Why?" repeatedly until you get an answer that reveals something that can be changed that you can also control. Fix that and you solve the problem and prevent its reoccurrence.

When we treat every failure as a learning opportunity, we strengthen both our systems and our culture of reliability.

Documentation and Record Keeping

Most maintenance requires knowledge of what to do, how to do it, and with what tools and parts. Maintainers do not keep all that in their heads, so documentation and recordkeeping are very helpful. Without the technical manuals, detailed plans and procedures, accurate parts lists, troubleshooting guides, wiring diagrams, and so on, they cannot do their jobs easily, safely, nor quickly.

All too often, some of that information is missing or out of date. The result is that maintainers end up guessing, or taking extra long because they need to find what's missing, often from sources that may be similar, but not exactly the same as the situation they are dealing with.

Here's a situation I've encountered: aligning a motor, gearbox, and compressor. If everything expands the same way as they heat up in operation, the task is relatively easy. But it can be complicated if one end of the compressor actually contracts due to cold suction temperatures and expands at the warmer discharge end. One end of the machine will expand up, while the other contracts and moves down.

Over time, changes are made in our systems: parts are replaced with better or cheaper ones, suppliers change, original manufacturers go out of business. If our documentation is out of date, we can easily make and repeat mistakes. Imagine if a design change by a maintenance engineer that improves performance isn't documented and the next time the equipment breaks down, the improvement is changed back to the original configuration. So much for all that effort.

Imagine troubleshooting an electrical system if the wiring diagram is inaccurate. You don't stand much chance of finding the problem. You'll try, but you'll be guessing and partway into your checks, you may realize that you made some wrong assumptions—then you'll start over.

Looking after documentation and keeping it current is a big job because there is a lot of equipment and there are all sorts of changes that can arise. You can have:

- **A dedicated clerk.** Have someone who takes care of keeping the documentation in order and up to date.
- **Online storage and access.** Keep documents electronically, ensuring that only the most recent version is available.

- **Procedures.** Any change to parts, materials, equipment, system configuration, drawings, and so on, must be fully documented so that the relevant follow-up changes are done, such as:
  - updates to online documents and parts lists
  - changes to the spares stocked in your storeroom
  - changes to procedures when equipment changes

In one mine, a process filter that regularly plugged was replaced by maintainers with a duplex-style filter that can be changed on the run. It was intended to help operators keep the process running without interruption. The change wasn't documented, and operators didn't know how to use it. Whenever it plugged, rather than simply moving the handle and informing maintenance, they would ask for the filters to be changed. No downtime was avoided. To make matters worse, the maintainers found that the old filter cannisters in stores didn't fit the new duplex-style housing.

Some work is required by regulations. There must be records that show proof it was done and often what was found or recorded. Usually, the regulations are related to something like safety (fire protection systems) or environmental protections (mechanical integrity of containment tanks and piping). There can also be industry-specific regulations like handling of radioactive materials, food cleanliness, or operating of hoists. Without proof that you did the necessary work and what you found, you can suffer fines or suspension of licenses to operate. Both are bad for business.

Many maintenance departments are poor at managing documentation. They often get the regulatory records right but fail to make their own work easier by keeping all the documentation up to date.

Accurate documentation doesn't just ensure compliance; it enables learning and process improvement.

Knowledge Sharing

Each time a maintenance job is carried out, there is a chance to learn. There may be minor tweaks to a procedure to make it easier to follow or improve the results. There might be problems spotted that could be corrected if people know about them. A maintainer might make a change that solves a problem, but if it isn't recorded, the solution will very likely be undone the next time it is worked on. If a part number is incorrect, it can be corrected. If a certain tool or part is found to work better than another, sharing that knowledge can improve the work for someone in the future.

It isn't practical for maintainers to share all their findings in meetings, so we gather that information in the form of text on work orders. If a problem that's been happening over and over again gets solved, we can make a note of it and how we achieve the solution. Then we can tell others in our company, perhaps even at other locations, about what we did, so that they too can benefit from what we learned.

Communication Is Key to Knowledge Sharing, and It Must Be Clear

If we are asked to do something and we fix it, it's useful to say what we found. For example, our work order tells us, "Equipment broken, please fix." Once the job is done, it's far more useful to record what was found and done than to simply write "Fixed."

Knowledge sharing, no matter how trivial, contributes to that continual improvement journey known as "excellence"! It is part of the basis for continuous improvement and cultural change explored in later chapters.

Today, digital maintenance systems make knowledge capture easier—but only if we use them well. Shared learning has the power to transform maintenance from a series of isolated repairs into an organizational capability.

Standardization

Standardization reduces variability, simplifies training, and lowers life-cycle costs—essential foundations for a reliable plant. Standardization can add to capital costs, but without it we add considerably to life-cycle costs—often much more than we saved at the project phase.

Maintaining multiple equipment variants increases complexity and errors. Each variation requires separate training, spares, and documentation, all of which add cost and risk.

Standardization gets us away from all those problems and added costs. To maintainers and operators, it is an obvious fix for problems they see every day. Maintainers influence, but don't control standardization, and that can lead to tension with engineers.

Standardization must be considered at the design stages for new systems. Often, the projects are budget-constrained, so they buy the lowest-cost items that will do the job. That usually results in a cheaper project but adds substantially to the life-cycle cost of operating and maintaining it later.

The solution is for your company to do a better job of preparing projects for operations and maintenance and before that by designing with standardization in mind. The decision criteria for whether or not to standardize needs to be based on life-cycle costs of the new systems and their impacts on other systems. What may seem more expensive up front to obtain a standardized item may actually be less expensive to the business because training and parts for it are already in place.

Like good information management, standardization forms another basis for continuous improvement and cultural change explored in later chapters.

Proactive maintenance practices form the foundation for *reliability and cost control.* When applied thoughtfully—and supported by accurate data, consistent documentation, and shared knowledge—they enable the cultural and strategic excellence we'll explore in later chapters.

PART II

MAINTENANCE MANAGEMENT FOR PERFORMANCE

Maintenance is a complex function that entails a number of activities all happening in parallel and many of them being actual work on equipment. A lot of work happens every day, and a lot of interactions make it happen smoothly.

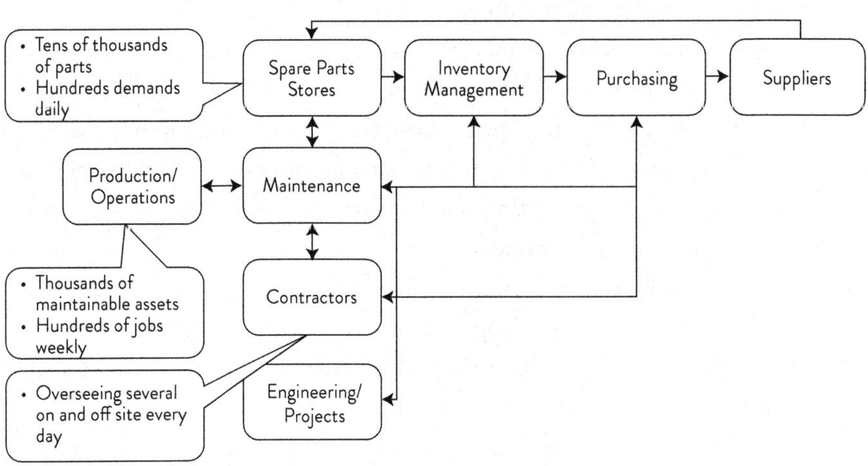

FIGURE P2-1 Maintenance Interactions

The basic or core practices used by maintainers remain the same, although the specifics vary from trade to trade, and the combination of trade skills required can vary from job to job. Getting the urgent and important work done first, making sure proactive work is done in a timely manner, running the background processes such as provision of parts and contractor labor, and making sure the proactive program is actually relevant is a major management challenge.

The main business process is known as "work management," as described in Chapter 4. It includes subprocesses like work identification, prioritization, planning, backlog management, scheduling. The process is usually managed with the help of computerized management systems that are quite complex and difficult to set up properly.

Doing the right maintenance isn't just a simple process of transferring technical manual content into a schedule. Understanding of reliability and how to manage consequences with various proactive maintenance techniques is needed to get the optimum mix of preventive, predictive, detective, and run-to-failure strategies.

People and skills are critically important as outlined in Part I. It's on management to ensure they have the right people with the right skills, and to provide the needed training and education to close any gaps. It begins with identification of needed competencies, a gap analysis to identify what competencies need to be completed, and then provision of the necessary means to close the gaps. Just doing training for the sake of doing training is not enough.

Like any business function, maintenance can be measured and improved. Knowing what you can influence with the metrics and how important it is to having meaningful metrics and "key" performance indicators. With literally a hundred plus possible metrics, it is easy to expend a lot of effort on data gathering, measuring, analyzing, and presenting. What's not so easy is to determine the metrics that really matter in your business!

One metric we all have an interest in is cost. How much are we spending? Is it too much? Too little? Are we spending on the right activities? Businesses love budgets and compliance is considered very important, but what if an opportunity arises to save money or generate more revenue, and the funding to do it isn't budgeted?

In this part of the book, I'll make reference to my other book, *Uptime*,[1] which is intended primarily for practitioners in maintenance and reliability. This book, *Steadfast*, assumes the reader is in a leadership role and deals with essential aspects of putting the concepts of *Uptime* into practice and sustaining them for the long term.

CHAPTER 4

PLANNING AND SCHEDULING FOR RELIABILITY

Planning and scheduling are two primary components in a larger work management process, depicted in Figure 4-1, showing interactions with the backlog, saved job plans, and the resources that will be needed to get work done:

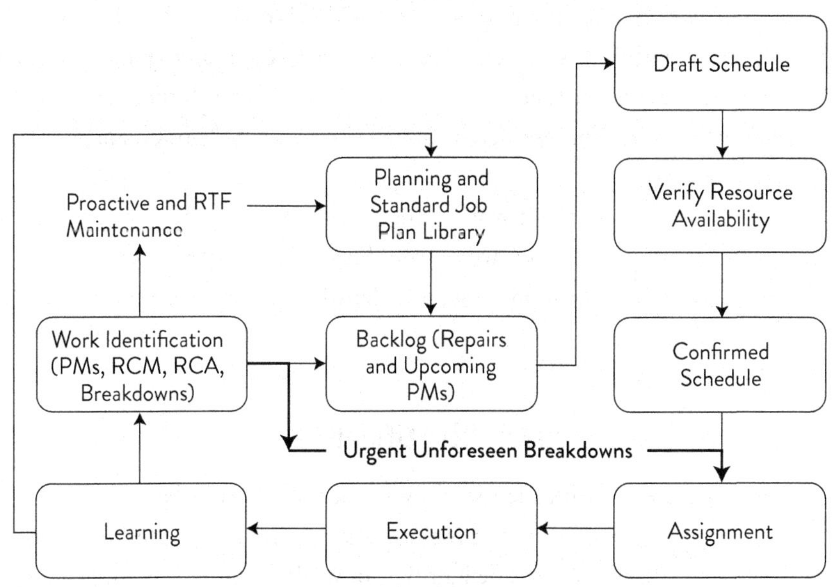

FIGURE 4-1 Simplified Work Management

This process starts at "work identification" and appears to be a simple once-through cycle for each job that arises—it is not. For instance, anyone should be empowered to identify work. That is a cultural element of reliability.

There is a buffer (or bucket) of work called "backlog." Planning shouldn't be done only after the work has been identified as needed. It can be done well in advance and saved in a planned-job library or database. Those plans are pulled as needed. Scheduling is a batch process. There are handoffs from one step to the next and plenty of opportunity for mistakes to be made—usually at the handoffs.

What are not shown are some important subprocesses that are crucial to those steps working as intended. Work identification takes several forms. Prioritization is thought to be quite simple, but in practice it is often done poorly. Planning is a whole process that can be supported with generative artificial intelligence (GenAI) tools. Scheduling is done weekly, not work order by work order. Between planning and scheduling, there is an important "staging" step where we make sure all needed resources either are in hand or will be available when the work is to be scheduled. Learning is often overlooked; most organizations merely close work orders without any learning.

Most organizations use one or another of the many available computerized maintenance work management systems (CMMS)[2] to aid in the flow of work and information within this process. While the use of computer systems is often taught to those using them, there is a widespread failure to teach those users about the underlying process. For instance, a planner should know "work management," not just how to plan.

Planning is a subprocess of work management. Showing the details of how planning, and other subprocesses work, would result in a very complex and likely confusing diagram. Those subprocesses are identified as boxes in Figure 4-1 and discussed here in the text.

Work Identification

There are two types of work identified by when they occur relative to a failure event and its consequences: proactive (before) and reactive (after). Reactive work is always "repair" work, and it is the default type of work, regardless of failure consequences. When something breaks, it is telling us it needs to be fixed. Considering that we know everything can break, it is quite possible to forecast all the possible repairs that will arise in the useful life of any asset. For assets that are important to your operations (critical), it's worth our while to do that. But for lower critical assets where we know the failure consequences are minor, we wait for them to occur and deal with them as they arise. Table 4-1 summarizes how to deal with critical and non-critical assets when it comes to work identification and prioritization.

TABLE 4-1 Work Identification for Critical and Noncritical Assets

Critical Assets	Noncritical Assets
Use RCM to define work.	Use existing PM or optimized PM, or allow to fail.
Use PM to reduce failure consequences.	Use PM only if cost justified.
Treat more urgently on schedules.	Work is not urgent.

The types of work are:

- **Proactive.** Preventive, predictive, detective—this is the least expensive work you can do. This is done where technically possible and where its costs are justified by the consequences of failure you'll be avoiding.
- **Proactive repair.** Correct failures before they happen—identified through predictive and detective maintenance. This is a repair to something that is failing but not failed. It is less expensive than reactive repair after it breaks. For this equipment you are not avoiding a repair, you are reducing its scope and eliminating the other consequences of the failure you avoided.
- **Reactive repair.** Equipment has broken down and is repaired—the most expensive type of repair. Costs can be reduced if the work is well planned and executed without urgency on a schedule. Equipment is allowed to break down if you can tolerate the consequences associated with the failure.

Anyone in your operation should be empowered to identify work by creating a notification or work request, either by phone, online, or through a supervisor. They will always be identifying defects or failures, and all of those people need to know your process for identifying work and how to use the CMMS or other tools to do so.

We'll deal with how to determine which of the proactive approaches (preventive, predictive, or detective) to use, in the next chapter. For now, assume those have been identified. Each of those will (or should) have a definition of actual work to be done (job scope), a frequency at which it should be done (e.g., daily, weekly, monthly), and we should know who (which skilled trade) is needed to do it. Once loaded into your CMMS, those should trigger automated work order generation at the predetermined frequencies, and they, in turn, should be loaded automatically into your draft schedule for work to be executed. Parts or other materials, tools, and so on that are required for those jobs should be stocked or known to be readily available (e.g., vendor management inventory in a workshop). If you cannot depend on the availability of those resources, you should be checking for them first and then scheduling.

Screening, Backlog, and Prioritization

Screening of new work requests is your first line of defense against duplicate work requests being raised. Duplicates often arise from two sources: multiple people all spotting the same problem and submitting a work request or notification, or from automatically generated proactive work orders.

Originators of work requests should be able to check easily if a problem has already been identified. One way is to put a tag on the equipment whenever work is identified. This can be highly effective and it is simple to implement. Another is to have it appear on screen when someone enters an equipment identifier—of course, everyone needs access for this to work. What many do is pass the notifications through a screening process by the area supervisor or planner. Screened work that is deemed to be needed is "accepted" or "approved." Whether it is approved or not, originators should always be notified so they aren't left wondering and possibly creating more requests for the same work.

How you've set up your automated work-order generation for proactive maintenance (PM) matters. When a proactive work order is due to be created, it can be added to the schedule, but it could be a duplicate if the previous work order for the same job wasn't yet marked as "complete." They can pile up fast, too. To avoid duplicate PMs, generate new PM work orders only "after completion" of the previous one. However, you need to be disciplined in closing those completed jobs quickly, or you could be causing PMs to be missed, leading to the failures you should have been preventing.

Backlog

All identified work that isn't on a schedule resides in a backlog. That's just a big bucket (usually computerized) of work to be done. We pull jobs from that backlog when we want to do them. Note that backlog grows with each new job identified, and it can grow excessively with duplicate work requests and auto-generated PMs.

Avoiding duplicates and managing how your automated scheduling handles new proactive work orders are keys to keeping backlog under control.

How Much Backlog Is Good?

With too much work backlogged, it becomes obvious that you are short-staffed. With no backlog, it probably means you have too many people for the actual workload. Under those circumstances, you'll notice that jobs start taking longer,[3] and even small jobs start to be delayed. A healthy level of backlog is two weeks of work per trade. For example, if I've got 10 millwrights each working 40 hours a week, then a healthy level of backlog for them is 800 hours of work. Needless to say, to make that determination, all jobs in backlog should have estimated levels of effort. Without that, backlog management is futile.

You can't hire and fire people monthly each time you do a backlog analysis, but if the average over a period of a quarter or half a year is used, you'll get a good idea of workforce and workload trends.

If backlog is growing steadily and you are doing what you can to avoid duplicates, then it may be telling you about a reliability problem. Too many breakdowns will quickly swell the backlog with lengthy repair work orders. It may be indicative that your proactive program isn't being followed or that it isn't effective. In any event, it bears investigation.

Backlog aging is an excellent leading indicator, and analysis of what backlog reveals helps us manage our workforce effectively.

Priority

Priority is assigned when jobs go in to the backlog, normally by a production or operations supervisor or a maintenance supervisor. Which jobs come out of the backlog first depends on the priority assigned to the job. Priority is an indicator of how urgently the work should be addressed. A typical priority scale is shown in Table 4-2.

TABLE 4-2 Sample of Work Priorities

Priority	Description	Response Time	Scheduling Implications
1	Immediate (as soon as possible)	Do work right away with no delays.	Breaks into ongoing work; always disruptive to established schedule.
2	Urgent (as soon as practical)	Do work as soon as workers become available.	Breaks into current schedule; disruptive, but it may be possible to avoid week-to-week implications.
3	Normal	Schedule work in next cycle base on asset criticality. For PMs, this work can be scheduled many cycles in advance.	Higher-criticality assets are worked on first, lower later (unless used to fill schedule gaps). PMs are done before proactive repairs.
4	Low urgency	Schedule within the next two to three cycles.	Scheduled by criticality with no rush. Can be shuffled to accommodate higher priority work.
5	Not urgent	Schedule work in shutdown or next convenient window that will not disrupt operations.	Usually schedules far in advance in window where production will not be disrupted by addition of this work.

Priorities are determined largely by circumstances at the time the work arises, and it is possible that circumstances can change and so too the priority. Note that work with priorities 1 and 2 are never scheduled—they arise in the moment and must be dealt with. The priorities for these work orders are usually added to the work order history after the work is completed.

Some people might call these "emergencies," but it is best to avoid that term as it implies some serious safety consequence that may or may not exist. You don't want someone to misunderstand and proactively call an ambulance when the job is really urgent because of production losses and not safety.

One thing to watch for with priorities is that they can be abused. If someone wants a job done quickly, they might ask for priority 1 or 2 even though it is not truly urgent. If you have a lot of urgent work, it can mean that your proactive program isn't working, your operations supervisors are abusing the system, the workload is very high, or you are understaffed—in those situations, everything becomes urgent. And, when everything is urgent, nothing is.

Planning

Planning[4] is used to shorten the job execution time by as much as possible while ensuring work quality.

Plans define what work is to be done; how (procedure, step-by-step instructions); to what standards; with what permits, parts, tools, test equipment, lifting and transport apparatus, consumables (e.g., lubricants); with what skills sets (e.g., mechanic, electrician); how many of each; and the duration of the job (from start to finish of the actual work). It may reference instructions in a manual, or technical drawings, wiring diagrams, and so on.

Unplanned work generally takes as much as three times or longer to execute than if the job is planned properly. That simple truth is the heart of the cost-savings business case for improvements in maintenance work management. One mine manager told me that unplanned work in their underground operations can easily cost as much as 14 times the cost of a proactive planned repair.

I use a planning game in my training classes. Teams put together Lego toys: One team has all the parts and instructions, another has some of the instructions and a few parts missing, the third has no instructions and a few parts missing. Time and again, the first team outperforms the second by a factor of three, and the third team by a factor of four or more.

If most of your work is unplanned, probably because it is reactive to breakdowns, your maintenance costs will be far higher than they need to be.

Given the big time and cost savings from planned work execution, and the productivity boost in your skilled trades workforce, it is worth noting that it doesn't take costly systems nor hiring to achieve. Making the shift to planned work, parts staging, scheduling, and disciplined execution requires no capital investment. It's all about implementing weekly schedule discipline, planning for recurring jobs starting with the most frequently arising, then kitting and material staging for those jobs. Those practices will create additional workforce capacity without any additional headcount.

Note that creating a perfect plan is all but impossible, and it takes far too long to do. We can create good plans relatively quickly, and planners should strive for that. Doing a job that is planned to 80 percent accuracy, is far better than having no plan at all. Allow feedback from the field to identify any needed corrections or refinements.

Proactive Planning

If you know something can break, you can plan for its repair. You do not have to wait for the breakdown before you plan it. We also have the means to save "job plan documents": file cabinets if you're old school, standard job plan libraries (by various names) in your CMMS. When work is identified, it is quicker to attach a saved job plan to the work order than it is to create a new one, but you can only do that if you have saved the plan. It may be in the form of an old work order that is easy to find or pulled from a databank or library and attached.

The best planners work on plans proactively. They do not work on plans for this or next week—for those it's too late. They build libraries of plans for use as above.

Most of your work should be identified long before it actually has to be done. All proactive maintenance is predefined and can be planned long in advance. If that proactive work can identify other work (e.g., vibration analysis that identifies the need to change bearings), then the work that can arise is also identified and can be planned. In combination, those two types of work should account for about 70 to 80 percent of your workload.

We'll discuss RCM later, but if you have used it, you will have plenty of "run to failure" jobs, where you've decided it is OK to let something fail because consequences of the failure are such that we can live with them. Those can all be planned in advance. In fact, if you've done a good job of proactively identifying work, you should have relatively few surprise breakdowns that were completely unexpected and for which you have no advance preparations.

Planning on the Fly

If there is no plan at the start of a job, as often happens with urgent jobs, the trade person must figure out what to do and how on their own. They will be doing it in the field without a lot of reference material, and they are likely to do it "on the fly" as they go, rather than all at once. Consequently, they will make multiple trips from the job to the stores, shop, tool crib, technical library, and so on that could have been avoided. They'll ask for lifting equipment that may not be available when they need it, causing a delay.

In most organizations, trades perform a quick prejob safety check to identify risks. To do that, they need to write down the major steps they will be performing to do the job.

By using a blank planning form (template), the trades can write down those steps, identify the risks, and mitigate actions they'll take. Add to that a list of the parts they need, tools, time it takes to do the job, and so on, and you have a job plan.

By creating the rough plan in the field, the trade organizes their work and can then work more efficiently. They don't like to waste their time, so they find this helpful. After the job is done, these plans are handed in with the work order and used to build your planned job library very quickly. If you are measuring the percentage of planned work executed, you wouldn't count this first job as planned, but you will see the metric climbing as those plans are reused for future work.

I often give my clients an editable blank template that they can use to start doing this right away—most use it.

Staging

There is no point in scheduling work if the needed parts, and so on are not available. You only want to schedule jobs that are "ready to schedule," meaning everything is known to be available or known to be arriving before it is needed to be used if we schedule within the next week.

A work order with plans should be listed in your backlog as "planned" but not "ready to schedule." To get to that ready state there needs to be a check and confirmation that all the resources identified in the plan are available. That can be done by a planner, a scheduler, a planner/scheduler if the jobs are combined, or a materials coordinator.

Note on organizational roles: Someone who may or may not have trades background can do this role. While experienced trades people will do it easily, someone without a trades background but with good organizational skills can do this very well.

They review plans and check inventory of stores, do physical checks if they can't trust their computerized stores records, gather parts into kits (or help the stores persons do this), and check on other resources such as rental equipment that may be needed. They do this for every job by priority that is "planned" but not yet on the schedule. Once the coordinator confirms that everything is available, the status of the work order can be changed to "ready to schedule."

Here's the sequence:

Identified → Planned → Ready to Schedule → Scheduled → Executed → Closed.

Scheduling

A work schedule tells us when the work orders will be done. Usually, it is best to schedule weekly and do daily updates to account for changing priorities or urgent work arising. Figure 4-2 shows how that process looks:

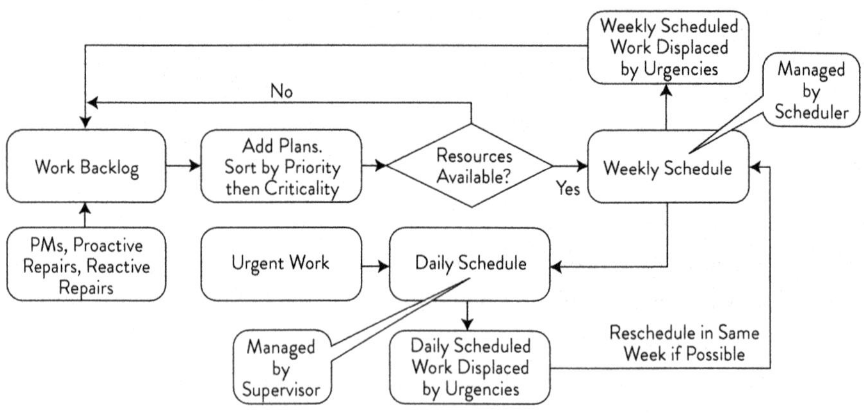

FIGURE 4-2 Scheduling.

Backlog jobs that are "ready to schedule" can be placed on a weekly schedule by trade.

Higher-priority jobs are loaded onto the schedule first, by priority, consuming all the capacity of the trade in that week. If there are too many jobs of the same priority, then equipment criticality[5] is used as a tiebreaker.

If the schedule has "holes" between scheduled jobs, then those holes can be plugged with a small job that may have low priorities. The objective of scheduling is to load up the full capacity of your trade workforce as much as possible.

Schedule jobs by trades to consume the capacity of trades available. Even if you know there are likely to be breakdowns that will interrupt it, schedule the full capacity, and allow your metrics around schedule compliance to reveal the problems you are having. Allowing for those disruptions is tantamount to allowing yourself to fail and not showing that you have a problem with breakdowns.

Do not attempt to schedule person by person. That takes far too long and it only takes someone's missed school bus to throw your schedule into chaos. Let the supervisors be the ones to figure who will do what day-to-day.

By its nature, you will be scheduling multiple jobs from the backlog for a week at a time. It is a batch process. Bear in mind that operational environments are very dynamic, and work priorities can shift at any time. Scheduling too far in advance is an invitation for disruption. Build the schedule for next week near the end of the week before. Get agreement on it with operations so they know you'll be taking equipment out of operation and when. They have a role to play in shutting it down, isolating it, draining it, and helping make it safe to work on. If you show up unexpectedly, you'll be throwing a huge delay into the job and ultimately disrupting the rest of the schedule. Each day, have a brief schedule review to deal with any work arising overnight or shifts in the priorities.

Schedule compliance is an excellent metric that speaks volumes about how well we are managing maintenance. In order to comply closely to your schedule, you need:

- Job plans
- Available parts, tools, and other resources as defined in the plan
- A schedule generated using the time estimates from the plan

- Disciplined supervision and execution to stick to job estimates and scheduled dates
- No, or minimal, disruptions to the schedule from urgent work. This speaks to the quality of your proactive maintenance program. The better it works, the fewer the disruptions.

It is also discussed in Chapter 7, but it is worth noting here. It should be measured on the basis of the first date you scheduled for the job after it was planned and scheduled—not the "required by" date from the originator. If you reschedule the work later, it may get done in a later week when rescheduled, but not when first scheduled. You can comply with the schedule for the week but miss it for the month or quarter when first scheduled.

Assignment and Execution

Once the schedule is fixed, it is in the hands of your supervisors for execution. They know the workload for the week from the schedule, and they choose who, from their crews, is best to do each job. They'll consider training opportunities for apprentices that arise from the work in hand. They'll decide who is best to teach them. They also see, daily, who's actually there and who didn't make it in to work. They have a brief meeting every day to assign the day's work to their crews. The crews then go about executing the jobs. Execution goes most smoothly with a good plan, everything available that is needed, and the right skills on the job. Having fewer disruptions mean faster completion and higher productivity. It also keeps the trades happy; most of them actually enjoy their work and dislike the disruptions and inefficiencies that can arise from poor or absent planning.

Learning

As work is being executed, any errors in a plan will become evident. If a needed part wasn't identified nor available, we need that feedback so the plan, or the staging can be improved. If an estimate of duration was way off, we need to know what it should have been. If the procedure identified a technical manual but not what specific pages, it could cost the technician time to find the needed information.

Sometimes a better way to do things than what is identified in the plan or procedure is discovered. All of those can be used to improve aspects of the work management cycle such as plans, staging activities, stores, documentation, and so on.

Continuous improvement depends on structured feedback from trades and supervisors, not just closing work orders.

CMMS, EAM, and ERP Systems

Managing maintenance entails keeping track of thousands of parts, procedures, standards, work requests, work orders, schedule details, estimates of work duration and effort, trade persons, and so on. Records of past work are useful in helping with future plans and to identify where reliability improvements are needed. Tracking it all requires structure and orderly record keeping. Before we had computers, this was all tracked on paper in file cabinets. Making improvements based on past experience required a great deal of time and effort. Since the 1980s, we have been doing that in specialized computer systems.

Purpose

We manage work (there's always a large volume of it) with computerized management systems. These systems are designed, some better than others, to support the maintenance function in the field. They are increasingly deployed over the cloud so they are always up to date with the latest version releases. It is important to realize that these systems are there to manage transactions and use the work order as an information-gathering mechanism. Work orders relate to work, not reliability, so these systems, despite the hype coming from their salespersons, are not designed for managing reliability. The simple reason is that work orders and failure modes are only indirectly related. One is not a proxy for the other.

Key Features

The CMMS manages your work management process and stores data about jobs (standard job plans) and work orders. It is set up with "master date" for work priorities, criticality in its asset register, units of measure, and so on. Within limits you can tailor (configure) it to match your desired work management and peripheral processes. Its database of work order with "status" indicators contains your backlog. The system will enable some calculations based on the data that it contains. Normally, those calculations are intended to provide outputs in the form of reports. They will usually come with many standard reports that are ready for use.

Information on parts used, time consumed on a job, start and completion dates/times, notes from the trades, records of reading they take as part of repairs, indication of what they actually did, the condition of the equipment, and so on are all normally available and associated with work orders job by job. Checklists and standard procedures are there for attachment to work orders as required.

Bills of materials, or access to them in other systems, are usually provided to assist in planning and execution. The system may contain a module for managing stores and even purchasing, but many of the systems are integrated with other management systems for those purposes. Enterprise asset management (EAM) and enterprise resource planning (ERP) systems tend to have all the functions integrated but are not always useful without considerable effort to configure them for use.

User Friendly, AI Enabled, and Mobile—It's the Way to Go

You will find that the more user friendly your systems are, the better. If your workforce uses mobile devices, it will enhance the efficiency of your work management process and provide you with far better field data than you will get with paper work orders.

Some systems are now coming with artificial intelligence tools to help you search old work orders, create job plans from your technical manuals or even from the internet, analyze work order history to identify proactive work that may not be effective, and identify where you have problematic equipment.

I've personally tried three of them, and they are miles ahead of the older CMMSs and EAM systems in terms of productivity-enhancing capabilities. Given the difficulty in finding and retaining skilled trades, those capabilities are needed more than ever.

The capabilities are growing all the time. Web-based systems that are mobile enabled are likely your best option and do seem to becoming much more prevalent in the market.

People like those systems with mobile capabilities, too. One customer trialed one with their trades, who made good use of it in the field. After the trial, they switched back to their old network system and the workers insisted on continuing the use of mobile capabilities. The newer web-based system had it, the old one didn't. After a taste of the new system, they simply didn't use the old paper-based work orders for writing things down anymore.

Your workforce these days are tech savvy and comfortable with it. They are uncomfortable with older paper-based ways of working and older tech that simply isn't convenient or easy to use.

Also consider how you support your systems. Information technology departments tend to get in the way with their valid concerns about security, cyberattacks, data integrity, who has access, and who can do what. They tend to default toward what they feel is "safe." That stalls the adoption of new tech and capabilities, and keeps your management systems cumbersome and inefficient.

Your workforce in the field simply won't use systems that are too challenging to use or user-hostile. Most of their work is with their hands-on tools. They are accustomed to the simple and easy-to-use tech in their personal lives. You will need to make your systems that easy to use if you expect them to be used. That trend can only be expected to continue. There's no going back.

Some points to remember:

- Train on the *process*, not just the software.
- Make systems user-friendly and mobile-ready.
- Begin leveraging AI in planning, scheduling, and interpretation of condition monitoring data.
- Align IT controls with usability.

Upgrading or Changing Management Software

This doesn't happen often, but when it does, it can be incredibly disruptive. Not many are installing a system for the first time anymore, but if you are, give it a lot of thought before acting. How you go about implementing a system (new or upgraded) will have a huge impact on how useful that system will be.

First of all, don't believe any of the sales pitch about business case that any software company will give you. The systems all add complexity, overhead, and costs. Setting them up requires a good deal of time and effort. All of that is expensive. The systems do automate some primary process like maintenance work management, spare parts stores integration with planning, and purchasing. But it only saves you money on those processes if you reengineer the processes first. If you simply automate what you do today, you may speed things up a bit but you won't add efficiency. You'll be paving the cow path, as a good friend likes to say.

The biggest benefit from any system comes from rethinking how you do things to make them more efficient and effective.

A presentation by Drew Troyer of Bootleg Advisors refers to work done by Troyer, Erik Brynjolfsson[6] of the Massachusetts Institute of Technology, revealing that by improving processes alone (without touching the technology) you can get a productivity boost; by focusing only on technology, you get a drop; and if you do both together you can get a substantial boost. That is amplified by Terry Wireman[7] in his books stressing the importance of process redesign before implementing a CMMS. Reliable Plant also urges process design first, then CMMS specification and selection.

Drew's presentation suggests that:

- Process-driven changes, with no technology driver, can deliver a *27 percent productivity boost.*
- Technology driven changes (upgrades, lift and shift), can actually *lower productivity by 7 percent.*
- A blended process and technology approach can *boost productivity by a whopping 75 percent.*

Many companies already have extensive investments in their management IT and are reluctant to make big changes. Those systems are continually being upgraded by their developers. If you upgrade to the next version, there is a strong temptation to do what they call a "technical implementation." You "lift" what you are doing now and "shift" it over to the new version with no changes. In theory, this results in a quicker implementation and less training on how to use the upgraded system. After all, you're teaching only new screens, not capabilities. But "lift and shift" intentionally avoids taking advantage of any new capabilities. I've got to ask: So where's the added benefit from all those features and functions that you won't use?

You can't trust that your IT folks will ever come back and do gradual upgrades in capability using the new features. Once they have a stable system operating, whether it is helpful to you or not, they will resist any effort to change it.

Management research and industry analyses consistently find that tech-only implementations underperform, compared to programs that blend technology with process redesign and workforce enablement. Those investments in technology that you are bound to make aren't needed only because your ERP supplier has produced a new version. They can actually deliver outsized returns, but only when paired with complementary changes in processes, organization, and skills. Brynjolfsson, Wireman, and others, including myself, have insisted consistently that the value is in the process and new ways of working, not the technology.

Why not invest a bit more in the process aspects of implementation first, as the experts are recommending, and then "go live" with a big gain that lasts?

CHAPTER 5

DOING THE RIGHT MAINTENANCE

Doing the right maintenance means selecting the most appropriate strategy for each failure mode—*before* it hurts safety, the environment, or production. In Part I, we focused on doing work the right way (efficiency). This chapter focuses on doing the right work (effectiveness): when to use condition-based, preventive, failure-finding, or run-to-failure approaches, and how to choose wisely.

Those choices involve two lenses: failure patterns (age-related, random, premature) and consequences. Together, they point to the right strategy for each situation.

Excellence is a journey that advances along two axes: efficiency (doing work the right way) and effectiveness (doing the right work). Figure 5-1 maps the main maintenance strategy options along those axes.

In the lower-left—no deliberate improvement—organizations drift into chaos. People lose confidence that change is possible and settle into "That's just how it is."

As the saying goes, "Whether you think you can or think you can't—you're right." Self-perception influences both our actions and the outcomes. If you are in that state where you think you can't, you have quite a challenge that is far more than just technical—it is in the hearts and minds of your workforce.

Keep circles of control and influence in view: what we control, what we influence, and what we can only monitor.

We'll discuss change later in the book. For now, let's just consider that it will be a big factor in how successful you are and what you can achieve by changing both how you work (last chapter) and what you do (this chapter).

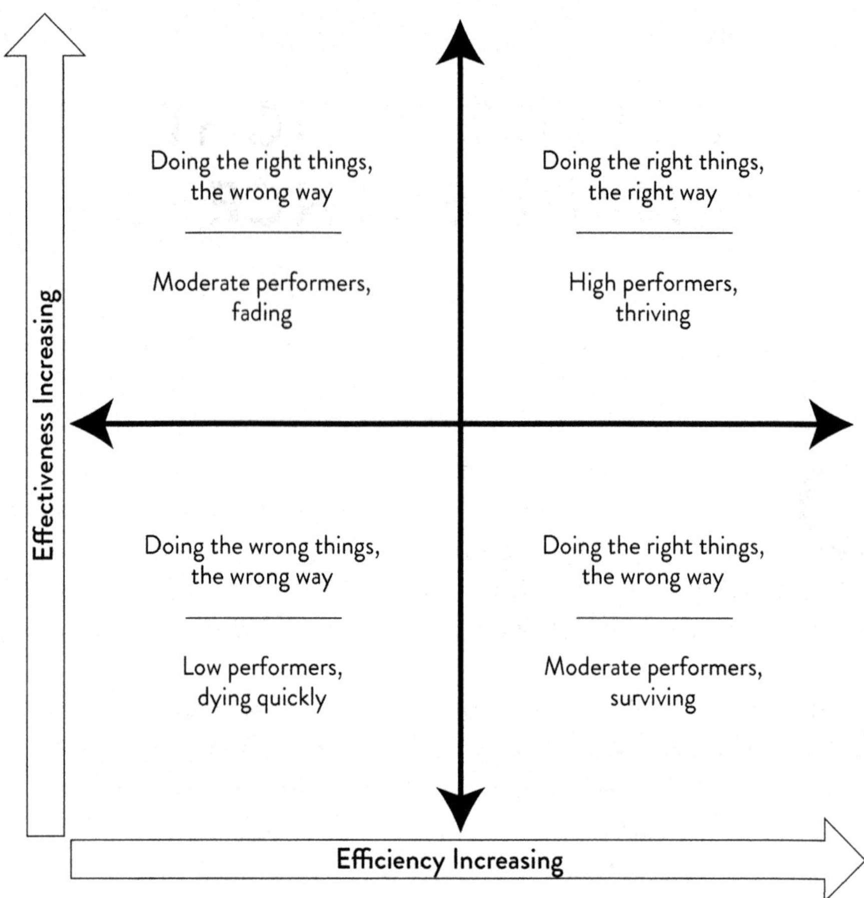

FIGURE 5-1 Efficiency and Effectiveness

Reactive and Proactive Maintenance

Reactive maintenance—waiting for a breakdown—becomes the default when we don't actively manage work. It is consistently more expensive than planned work. Many experts and my own experience will tell you that it is at least three times more expensive. And yes, some failures are not worth preventing—but the point is to know which ones, not to be reactive by default.

Proactive maintenance means deciding ahead of time what to do, why, and when—including when not to act. Selection requires some analysis: reliability-centered maintenance (RCM), proactive maintenance optimization (PMO), root cause failure analysis (RCFA), followed by disciplined follow-through.

Proactive maintenance is all about taking maintenance action before you suffer the consequences of failures. In some cases, we actually do wait for the failure to happen before acting and accept a small risk that we'll be caught short, knowing the risks and consequences of that failure. In other cases, we wait for the failure to start happening before we take action, and then we act quickly to avoid consequences, knowing that we've actually taken full advantage of the useful life of the asset between failures. In some other cases, we know not only that the failure will happen, but also know with some confidence when (age, or some other measurable gauge like hours of operation or quantity of production). In those cases, we act before that "when" and replace or restore the item before it fails. That's the only case in which we actually avoid the failure entirely. Finally, there are some cases where we can act, but the cost of doing so is not justified by the cost consequences of taking no action—we let those run to failure.

Consequences

Every failure has consequences in one or more of the following classes:

- Increased risk to health and safety
- Increased environmental emissions
- Minimal disruption requiring repair (a nuisance)
- Loss of redundancy but not production (important in utilities and highly dangerous operations)
- Loss of production
- Risk to company reputation
- Loss of market share
- Loss of regulatory approval to operate

We carry insurance to cover some of those risks, but a better form of insurance is to eliminate or at least mitigate the known risks. Why not be proactive and potentially enjoy lower insurance premiums because the risk is lowered? Insurance companies do understand that!

Precision Lubrication as a Reliability Enabler

One of the most basic things we can do for our equipment is to sustain good operating conditions. We can keep it from overheating and keep dirt from getting inside by keeping it and the area around the equipment clean. Just as we make sure the oil in our car is topped up, filtered and replaced periodically, we can also keep our industrial equipment properly lubricated.

Many organizations underestimate how many failures can be avoided by lubrication selection, application precision, contamination control, and sampling discipline. Precision lubrication is one of the most cost-effective reliability levers because it reduces failure frequency at the source: friction, wear, heat, and particle-driven damage.

Getting started with precision lubrication does not require heavy investment: define lubricant standards and consolidation rules; label and control storage and handling; apply contamination control (clean, dry, sealed); specify correct quantities and intervals; and implement representative oil sampling with clear alarm limits tied to specified actions. Many high performers opt to have a dedicated lubrication technician taking consistent care of all this. When these basics are in place, oil analysis becomes far more predictive and trustworthy. The analysis data will reflect the machine and lubricant condition, not poor sampling or dirty handling.

Done well, precision lubrication supports both preventive and predictive maintenance, improves safety and housekeeping, and increases the credibility of reliability programs because results become visible in reduced failures, cleaner oil, and longer component life.

Failure Patterns Matter

Some components wear out with age. Many other failures are randomly triggered while an asset is operating. A surprising number occur prematurely due to human error or latent defects. Knowing which you face is the first step to choosing the right strategy.

"Failure modes" are events that result in loss of use of the asset. They are events such as tire blown, fuse blown, shaft broken, belt snapped, or bearing seized. Each has a cause and those causes can be related to age, usage, or even random events.

In many complex systems, most failure modes are not age-related; a smaller fraction exhibit clear wear-out with age.

Some failures arise because of aging effects: wear, corrosion, erosion, and fatigue. The materials or their properties resistive to failure are gradually expended as they "age," as they are exposed to the conditions that cause deterioration (e.g., corrosion and erosion), or as they are used, as in wear and fatigue. If we know the age at which they become deteriorated to a condition that we can no longer use, we can restore or replace them using "preventive maintenance." Studies have shown that roughly 11 percent of the failure modes in complex systems occur with aging effects.[1]

Random failures don't just arise on their own, but they are triggered by random events. For example, a tire blowout on your car occurs randomly, while the wear-out of the tread occurs with usage. Rolling element bearings are actually designed to wear out very slowly (age) but usually fail randomly because of errors in lubrication, overloading, or other random events that initiate one or more of their various mechanisms of deterioration. The failure mechanisms are known, but when they will show up is random. Statistically, studies have shown that roughly 89 percent of all failure modes occur randomly.

A premature failure occurs long before the aging or even the random events that the items are normally exposed to can take place. They are random and occur extremely early in the component's life, usually very soon after installation. Defects in the materials that the component is made of, installation errors, manufacturing defects or errors, and even errors by operators as they start up the equipment, can lead to premature failures. We humans are far from infallible, and we all know that we make plenty of mistakes. Premature failures can usually be traced back to a human error of some sort. The studies that gave us the 11 percent and 89 percent figures earlier also show that 69 percent of failure modes occur prematurely. These failures can show up in testing after build or installation, and it is a good idea to also perform testing after each maintenance job. It is better to find these failures in test situations before handing over to operations.

Knowing this, manufacturers put a lot of effort into quality control and even testing before putting a product out for sale. Quality control will catch some of the errors, but not all. They also offer warranties on materials and defects for new products that you can buy. They'd rather make good on their (or their suppliers') errors than lose a customer. Only testing of newly manufactured, built, or installed systems can reveal most premature failures before the equipment enters service.

Regardless of how a failure is initiated, it requires repair. Early detection lets us time maintenance intervention to minimize consequences.

Condition-Based Maintenance

Most failures are initiated by randomly occurring causes. That means that we won't know when they will happen, but we can forecast the mechanisms of deterioration and failure that will ultimately result in the equipment breaking down after the failure is initiated. In some cases, failures happen very suddenly and randomly (e.g., electronic devices and even human errors), so there is nothing much we can do because we won't have time to act. We'll talk about those later. In most other cases the failure mechanism starts small (like a crack) and gets worse (eventually the crack becomes a break). Often, as the deterioration occurs, signs that it is happening can be detected if we are watching. We can find cracks if we inspect.

We can assess bearing condition by monitoring temperature, vibration, ultrasound, and lubricant condition. We can find power transmission equipment deterioration (e.g., belts and gears) by monitoring thermal signatures, vibrations, sound, and ultrasound. We can find electrical device breakdown by monitoring ultrasound and thermal emissions. We can find air, steam, gas leaks, and even some electrical faults using ultrasound. Many techniques and a lot of technology can help us do those monitoring activities to reveal failures that are already in progress, but not yet progressed to the totally failed state.

Operator Care—Low Tech, High Touch

We can also use our human senses. We can hear equipment in distress, see leaks and dirt build up, feel high temperatures if they get glowing hot, we can feel vibrations, we can smell lubricants that are too hot and giving off vapor. With simple tools (a screwdriver or a stethoscope), we can hear distress in bearings, air and steam leaks, slipping drive belts, and damaged gearboxes. We can even detect unusual "humming" in electrical equipment.

All of those human and technology-aided methods help us monitor conditions that are indicative of equipment in distress—those are all called "condition monitoring."

Detecting most of those (except perhaps cracks) requires that the equipment be running. One big benefit of condition monitoring is that most of it can be done without disrupting operations. However, when we find a distressed condition, we know that the equipment is now "failing" and that it won't be long before it fails. We need to act in time to correct whatever is causing the condition before that failure occurs, or suffer whatever consequences accompany the failure. Carrying out that follow-up "proactive" repair, after detecting the condition, is known as "condition-based maintenance."

Since condition monitoring can find a large number of harmful conditions, and it is not disruptive to operations, it is highly effective at enabling the reduction of failure consequences for our business. It works on many of the randomly occurring failures as well as some that occur with age or usage.

Car tires can be maintained both by replacing with age (e.g., after 60,000 to 80,000 km), or we can monitor their tread depth indicators to tell us they are near the end of their useful life. Let's say our driving style is somewhat gentle and we drive on good roads. We'll probably get closer to that 80,000 km point. If we used a usage-based replacement strategy at 60,000 km, we will be sacrificing 25 percent of the tire life. If we use the tread depth indicators, we'll get much closer to the 80,000 km and possibly even exceed it. Fewer tire changes mean lower maintenance and operating costs and fewer trips to the tire shop. Condition-based maintenance in that case is more cost effective.

Indeed, if it is technically feasible to use it, condition-based maintenance is often the superior choice.

Preventive Maintenance (PvM)

Some failures occur only as a result of aging effects. Elastomeric compounds are always "curing" and getting harder. Their use in equipment depends on them remaining "softer." As they age, they become more brittle and susceptible to failing (usually leaking). Other components wear out because of material loss due to friction (e.g., materials wearing on belts or chutes, wear of mill liners that are used to pulverize rocks in mining operations). Others suffer fatigue as a result of repeated stress cycles. Some erode due to flow of water or other liquids, and some corrode due to chemical interaction of materials and fluids. These all happen at known "rates," and since we know how much material there is to begin with, we can determine the age at which they will be worn, eroded, corroded, or fatigued to a point where they have little to no more useful "life" (material) remaining. At that point we can restore them or replace them. That is known as "preventive maintenance": replacing or restoring at a defined age or usage to avoid failure.

To be effective, we need to know how long the items have been in service since last replaced—either elapsed time, operating time, or some other metric that tells us it is worn, eroded, corroded or fatigued.

We might use age, running hours, tons of product transported or passing, or operating cycles. All of those can be metered, triggering our restoration or replacement activity when we reach the predetermined metered limit. There is no need to inspect the condition, because we know the item's usefulness has been expended.

Some of those failure mechanisms also allow us to monitor condition by inspection. We can see and measure wear, material loss due to erosion, and extent and depth of corrosion. We can't see fatigue, but we might be able to detect cracks as they form and propagate.

Where age estimates are uncertain, we prefer condition monitoring to capture items aging faster (or slower) than expected. It's quite likely that operating conditions have not been entirely uniform, so the actual age at which replacement is needed could arise sooner or later than our calculated limit. Condition monitoring, where practical, can catch those items that age rapidly and also those that will last longer.

Detective Maintenance— Failure-Finding (Functional) Testing

Where high reliability is wanted, a lot of design effort goes into component reliability and design. The most common way to ensure a system continues to function is to make it somewhat fault tolerant, and that is usually achieved with some level of redundancy. Consequently, we often run into backup or standby equipment and devices that under normal circumstances do nothing. When the item they are backing up fails, they come into use so that the systems' functions are not lost. We see a lot of that in systems where reliability really matters: aircraft; nuclear power generation; electrical, gas, and water utilities; and telecommunications. The backups and standbys are normally dormant. Because they are dormant, they may have different failure modes that occur only when the equipment is not in use. To ensure they will work as intended when we need them, we can test their operation periodically. That testing is known as "failure-finding testing": we are detecting failed states after they've occurred but before we need the item to be used. In doing the testing, we are reducing the risk of it being unavailable when we may need it.

There are also systems and devices that are installed to protect us and equipment from dangerous conditions that can arise when failures occur or when external situations arise. These devices, like backup equipment, provide a protective function. They don't necessarily provide backup functionality, but they lower the consequences of the situation by acting.

Such devices include:

- Alarms
- Trips
- Shutdowns
- Warning signs
- Exit signs and evacuation route markers
- Fuses and breakers
- Parachutes and life vests
- Personal protective equipment
- Fire alarm and suppression systems
- Explosive gas and radiation detectors
- Secondary containment structures, guards, and barriers

These and many other devices like them can be found anywhere! We rely on these devices, heavily at times, yet we tend to forget about them until they are needed.

All of these have functions that require them to act or be capable of acting when something else goes wrong. They need to be available. And they can fail while they are in their dormant state waiting to be used or actuated by that other event they are protecting against. Some cannot be tested without using them—for instance, you can't test a parachute before you use it, so you take extra care in packing it. In fact, the best person to pack a parachute is the person who will use it!

Fortunately, most can be tested. Sometimes, the test can be automated as it is in cars, but not always. The testing proves that the item worked up to and at the time of the test, but doesn't guarantee it will work after the test is over. There is a probability that the device won't work after it has been tested but that probability is normally low. By testing frequently, we reduce the time between tests and thus reduce the time to which we are exposed to that low probability that it won't work. We don't guarantee it will work, but we do reduce the cumulative probability that it won't work during that time between tests.

Running to Failure (RTF)

As discussed earlier, RTF is absolutely fine if we can live with the consequences of that failure. Not all failures have serious consequences. If it costs more to prevent a failure than the consequences that occur when the failure happens, then why not simply do nothing and allow the failure to occur? If we make that decision consciously, we can also plan for the repair and have necessary parts and such available. RTF doesn't mean being inefficient when we repair it!

RTF is not neglect: it requires spares, access, and a planned response.

Choosing Wisely: Picking the Right Maintenance

Knowing that everything can fail, understanding how it is used and the failure mechanisms that can occur, knowing whether they occur with age or usage, randomly or prematurely, and knowing the consequences of its failure enables us to make some smart choices.

If the failure provides some early warning that we can detect, then condition monitoring would be appropriate. This applies to most randomly occurring failures except those in electronic devices, which give off very little warning at all.

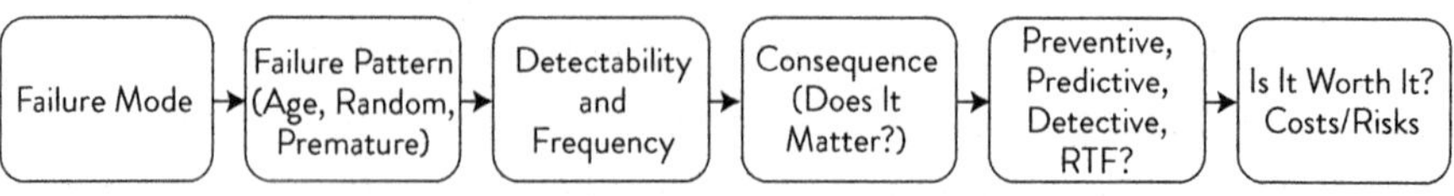

FIGURE 5-2 Simplified Logic for Choosing the Right Maintenance

If a failure occurs at a more or less consistent age (or level of usage), then preventive replacement or restoration would be appropriate. If that same age or usage failure gives some early warning, it may even be a good candidate for condition monitoring to stretch its useful life.

We know that human error occurs randomly, without any warning, and is pretty much impossible to predict or prevent. Most humans don't make mistakes on purpose, there is always a cause—inattention, fatigue, procedures that include mistakes, lack of procedures, lack of standards, lack of training, ergonomics that make it difficult to do something well, lack of checks on work that is performed, errors because they are rushed, and so on. If we know that human error is the cause, we can usually change the circumstances that give rise to it: procedures, training, enough rest, checklists, and so on.

If we know the failure is more likely to happen prematurely, we would be wise to avoid disturbing the item except to deal with its ultimate failure. Take care in purchasing, constructing, building, commissioning, using, and maintaining the equipment so as to make as few mistakes as possible. I've dealt with many failure situations that can be traced back to initial installation errors that can be very difficult for maintainers to correct once the operation is commissioned.

For protective devices, testing is appropriate unless it is impossible. In that case we must either live without it, find a way to make its dormant failure apparent, or design in redundancy.

A parachute is a good example: The jumper packs it and has a secondary parachute. We can't test air bags in our cars, but we do have many of them. If one fails, others will work.

Sometimes, genuine design flaws occur in our systems and equipment that give rise to failures that are either difficult or impossible to prevent, predict, detect, or otherwise avoid. If those have significant consequences, we can consider changes to the design itself. Those changes tend to be expensive and challenging to implement though, so they are not always implemented in systems that are already in service. The best time to spot those is during design.

Finally, if the failure has no significant consequence and we can live with it happening, then consider running it to failure and being prepared for the repair when that eventually (and inevitably) occurs.

Methods to Help You Choose

Doing nothing is a choice—a bad one—resulting in high costs and a lot of chaos.

Choosing to do the right maintenance can be done with the following methods, often in combination: reliability-centered maintenance (RCM), PMO, and RCFA.

Reliability-Centered Maintenance (RCM)

RCM was developed in the late 1970s in the aircraft industry and has become the "gold standard" in selection methods for choosing failure management strategies. It has resulted in significantly huge improvements in aircraft safety and maintenance costs. It has been deployed in aircraft, nuclear power, utilities, and many other industries where reliability is truly important.

To learn how to do it, you can look into *Uptime*[2] or another book I've written, *Reliability-Centered Maintenance—Reengineered.*[3]

RCM is sensitive to the operating context, so it accounts for differences in how an asset is used from instance to instance. It looks at the desired performance of each function, the possible failed states, and what causes them, then uses a structured decision logic to determine the most appropriate failure management strategy, considering both technical feasibility of the option being considered and whether or not it is worth doing in that context.

RCM is very thorough and somewhat time-consuming. I have personally used it since the 1980s and have always found it to produce excellent results and performance in the systems analyzed. Operators and maintainers learn a lot about their systems that they didn't know before. They participate in the decision process, so they take ownership of the results, and if those results are applied, the analyzed systems' performance has always improved. It has consistently lowered costs and improved asset availability.

I have done many such analyses and found that most can be carried out within a week using a team of three or four personnel. They need training to do it correctly, but that's only about two days and can be done in a variety of ways that are not disruptive to your operations. Although the demands are not huge, to some organizations it is difficult to make those people available. Because of that, it is not recommended for systems or equipment that are known to have relatively low consequences associated with their failures. I recommend it (strongly) for anything deemed critical because of operational, safety, environmental, regulatory, or reputational reasons.

RCM is for your most critical systems and assets. Do not attempt it without training and without some help initially until you get good at it. Many have tried that to save money and ended up making mistakes, either making things worse for themselves or giving up on it entirely. Neither of those is desirable. RCM is worth doing and as such it is worth doing right.

Proactive Maintenance Optimization (PMO)

Like many organizations you may not have done RCM, but you do have a proactive maintenance program. It could be a result of simply following manufacturer recommendations or other methods. Those manufacturer recommendations are often flawed. The manufacturer does not know your operating context, and it usually matters quite a lot. The manufacturer also very seldom, if ever, uses the equipment they sell to you, so their actual experience with it is quite limited. Do they really know how it behaves in service in your context? Have they experienced how it fails and what to do to correct, prevent, or predict it? Do they sell you the parts you use if you follow their recommended maintenance (which is often very heavy on preventive replacements)? Does their warranty insist that you follow their recommendations or risk voiding its terms?

What I'm asking about isn't evidence of a conspiracy theory, but it is common practice. They have a vested interest in avoiding warranty claims, so they set conditions. It is not unreasonable to insist you follow their recommendations. But it is unreasonable to expect them to know your operating context, how failures will occur, and how they will matter to your operation. If they've not done something like RCM, they haven't been particularly thorough. Even if they did it, they would not know your operating context, so they can't possibly make decisions that are best for your operation. For example, a pump starting and stopping to occasionally drain a sump, will fail in different ways than that exact same pump running continuously pumping a highly corrosive and toxic liquid in a chemical process. Both pumps require different maintenance, but the manufacturer's instructions will not have taken that context into account.

The net of all that is that manufacturer recommendations are seldom optimal for your operation. If you are following them, you will have some experience that reveals how effective those recommendations have been (or not). You may have already modified some of them based on that experience.

You may also have some maintenance developed as a result of doing RCFA in the past. If you did that right and truly nailed it, those maintenance tasks are probably pretty valid. But if the program was developed by RCFA or manufacturer recommendations, and modified over time, it is still possible that your operating context has changed and some of the tasks are now suboptimal.

You can optimize those tasks using the logic described already. Look at each task and determine if it is dealing with something aging, random, premature, or protective. Is what you are doing technically feasible? And if so, is it worth doing it based on the consequences of not doing the task? Are you doing it at the most appropriate frequency? Can that be stretched out, or should it be shortened?

You can improve on an existing proactive program using that sort of logic. Even if you use RCM on your critical assets, you can optimize the program for the noncritical assets using this approach.

PMOs single biggest flaw is that you can't optimize if there is no PM to optimize. If there is a PM that is missing, it won't find one for you. You need to have a way of identifying failure modes before you can identify new tasks—that's what RCM does.

Use PMO when you already have a PM program and want to cut waste and add missing condition-based maintenance (CBM) or failure-finding where appropriate.

Root Cause Methods

Everything that happens, including failures, is the result of some other event or conditions. You can have fuel and oxygen together safely, until you add heat. You can run equipment hard, but if you exceed its limits, you could break it. You can operate and maintain in accordance with good procedure and practices, but a new untrained maintainer who doesn't know about those procedures can make a mistake.

Whenever a failure occurs, you can dig into the causes of that failure. Understand what went wrong—the events and consequences—dig into what could have happened, eliminate those that are not supported by the evidence that is available, and test the ones that are possible. One method for doing this that I like to use is called "Five Whys." It is extremely simple to use: Simply ask, "Why?" for each observation.

Here's an example from my own experience in petrochemicals. A furnace fuel pump suddenly stopped, causing a major plant upset. We didn't want it to happen again, so we began asking, "Why?" It stopped suddenly and without any warning. "Why?" The repair crew saw that the pump internals were severely damaged, and that was unusual for that pump—it pumps fuel that has some lubricant properties. Past history reveals only bearing failures, not internals. So why did that damage occur this time? We found that a new operator was working that day. The operating logs revealed that there was an upset in another process at the same time and the operator was away investigating. The operations data historian revealed that an actuator on the pump suction valve had been closed just prior to the failure. It had to be the operator who did that, and we asked, "Why?" again. He had mixed up the two systems, and in troubleshooting the other, he had closed this valve. That gave us a plausible root cause: lack of system knowledge on the part of a new operator. Either he hadn't been trained fully, or hadn't remembered the training. In either case, a new operator shouldn't have been out there on his own. The mistake was not his fault; his training and production supervision were both investigated further.

Investigations of failures often reveal operator or maintenance errors; human error in general is a huge factor in any complex operation and cannot be ignored.

Root cause methods are very accurate and deliver great results, one failure at a time and after the fact. However, you cannot develop a maintenance program efficiently with root cause methods. It will take far too long and you will suffer a lot of failures and their consequences as you do it.

PM programs are built from a variety of methods. The most effective is RCM. PM optimization is very good, but limited only to existing tasks, and root cause methods, while very effective, are slow and entirely reactive to failures as they occur.

What if I Have No PM?

Some operations don't actually do any PM. They are running everything to failure and then fixing it. It's inefficient, highly disruptive to operations, and very expensive to do that. There could be many reasons why they are that way, but underlying them all are a lack of proactive mindset and a high tolerance for chaos.

Eventually, that tolerance changes, and someone sparks the idea that being proactive can help them cut costs, improve on the chaos, and reduce production variability that it causes, and possibly even produce more. But they don't know where to start. Highly reactive organizations almost universally lack in their work management practices and have very poor technical documentation and record keeping. Chances are that even if they have a CMMS, they do not use it very well. They spend a lot of money on rush purchases and probably have a lot of parts stashed away all over the plant that are not on the books.

If there is technical documentation, you could go through it and look for the manufacturer's recommendations and then put them into practice, but consider the manufacturer's motives before trusting explicitly. It is usually better to look into the rapid deployment of various condition-monitoring technologies.

90-Day Quick-Start

1. Assign someone (an experienced maintainer or operator) for checking on oil, then topping it up, and keeping things greased. That alone will probably lower the chaos. If that person is diligent in using their senses to make observations, you'll start to hear about conditions that need to be dealt with. Expect to find leaks, hear rattling and squealing, and feel excessive heat and vibrations. They are identifying failures before they have reached that totally failed state. Encourage creation of work requests for all the observations. Have a supervisor and a part of your work crew verify them, and address them before they become failures. Recognize and reinforce for finds and for each person that they teach how to find the problems.
2. Operators can learn a lot of what often gets left to maintainers with very little training. Remember the high-touch but low-tech approach. Teach them to use their human sense to detect anomalies—anything they can notice that is unusual—and initiate a work request to fix it. If you avoid a big failure, recognize it and reinforce the behavior. Make it easy for your employees to want to do what's right, and encourage more of it.
3. Clean up messes and dirt from equipment, especially on cooling surfaces. While cleaning, watch for defects: small leaks, vibrations, excessive heat, and so on.

4. Get analysis equipment and train someone in its use, or hire a contractor for:
    - vibration analysis on rotating equipment (no less frequent than once a month)
    - thermal imaging (infra-red) checks on electrical equipment and drives such as belts and gearboxes that can get hot
    - oil sampling and lab testing for larger sumps of lubricants
    - ultrasonic scans looking for leaks and electrical faults

All of those contribute to a reduction in chaos and production variability, which leads to improved quality and quantity of production, as well as lowering your costs. Use some of those savings to reward those who make the saves. Encourage that proactive behavior.

Shifting to proactive thinking also means a shift away from encouraging reactive behaviors.

Stop encouraging late night repairs and callouts. Those who do come out will still get their overtime or callout premiums, but stop giving encouragement for it. Do not provide anything extra, because all that work means that failures were not caught in time.

Question whether those urgent callouts by operations are really imperative. Often, they treat any breakdown as an emergency even if already running on backups.

Before allowing overtime or callouts, ask if the repair is really needed so urgently. Understand the consequences of the failure well enough to make an informed decision. Is there a backup? Is it running? Is it likely to last until the next day or next week? Breakdowns are not always cause for urgency and all the added costs of being unprepared. If you decide to wait, explain your logic to your operators so they can learn.

As the chaos comes under control, introduce root cause analysis on those failures that keep cropping up and those that have big negative impacts.

Over time, your program can evolve and get better. Getting started is actually much easier than you might have thought.

Training

There are courses and consultants who can help you with root cause analysis methods, with Five Whys, with PM optimization, and lightweight screening to prioritize assets for full RCM.

For root cause and PM optimization, you'll need a solid understanding of the principles taught in RCM. Consider RCM training, even if it isn't your intent to use RCM initially.

The training for RCM speaks to all the considerations you need to make good decisions about managing failures and their consequences. It is applicable to your use of root cause methods like Five Whys and PMO. It informs your decisions on where to deploy those condition-monitoring techniques to get you out of that chaotic RTF state.

If you want to do RCM, you will need formal training and expert facilitation to get those first few analyses done. You may want to invest in your own in-house facilitators. Don't avoid the training though; I've seen many cases of disappointing results where they shortcut the process.

Attempting RCM analysis without the training or without facilitation turns out badly. The concepts are not complicated, but they are new and easily confused. Your maintainers are happy in the field but not with analysis work. They need good facilitation.

Once you decide to do RCM, do it right, don't cheap out.

Summary

For managers, here's a shortcut to the logic:

- What's the failure pattern? (age or usage related, random, premature)
- Can we detect it with enough warning to act? (yes → CBM; no → consider preventive, detective (testing) or RTF)
- What consequence? (health, safety, environment, production, compliance, reputation)
- Is the task technically feasible, and is it worth doing here? (Consider the evidence.)
- For human error: what must we fix to ensure it doesn't happen again?

For practitioners, Table 5-1 summarizes the various options and when to use them.

TABLE 5-1 Failures and Their Maintenance Strategies

Failure Characteristic	Primary Signal	Best Default Strategy	Why It Fits	Watch Out for . . .
Age-related (wear, corrosion, erosion, fatigue)	Predictable end of life (age, cycles)	Preventive maintenance (PvM) at regular intervals.	You know when these failures are likely to arise. Easy to schedule.	Variability in operating conditions can change predicted life. Consider CBM if possible.
Random (initiated by known stochastic causes)	Asset condition shows gradual deterioration initiated at any time.	CBM at regular frequent intervals, act on deteriorated condition.	Early identification reduces failure consequences without sacrificing useful life.	Human error and most electronic failures do not come with warning.
Premature (human error/ latent defects)	Occurs early in "life" of asset after installation or maintenance.	Error proofing + quality assurance and quality control ≠ testing before putting into service.	Prevents introducing faults and finds failures before the item enters service.	Requires disciplined execution, procedures, standards, checklists.
Hidden	Only in protective devices that are "dormant" during normal operation.	Detective maintenance (DM). Functional failure finding tests.	Reduces risks associated with multiple failure events in protected systems.	Testing interval-based tolerable risks.
Low-consequence failures	Can be treated as a nuisance.	Run-to-failure and plan repairs in advance.	Avoid overmaintaining where consequences are trivial.	Have plans and spares in place so nuisance failures don't grow into critical problems.

Closing: Turning Choices into Habits

Choosing the right maintenance isn't a onetime workshop—it's a habit you build into how the plant thinks and acts. The logic in this chapter gives you the vocabulary (age-related, random, premature) and the playbook (CBM, PvM, failure-finding, run-to-failure). The goal is simple: reduce consequences while spending effort where it actually pays.

Start where signal is strongest. If a failure pattern gives you detectable warning, lean into condition-based maintenance and act in time. Where age dominates, schedule restoration or replacement at a sensible interval—and verify with evidence so you're not throwing life away. For protective and dormant functions, treat failure-finding as nonnegotiable; it quietly buys down risk every day. And when consequences are truly small, plan to run to failure on purpose, with spares and access ready. That's not neglect; that's design.

The human side matters just as much. Premature failures come from our own processes—how we procure, install, lubricate, test, and hand back to operations. Error-proofing, checklists, post-maintenance testing, and operator care convert good intentions into reliable outcomes. Celebrate saves found early; learn quickly from the misses.

As your context shifts, let the program shift with it. Operating modes change, loads change, teams change. Revisit assumptions, test intervals, and task types. Use RCM on the few assets that can really hurt you; use PM optimization to tune the rest. Treat root-cause work as the feedback loop that keeps the whole system honest.

If you remember nothing else, remember this: Don't maintain everything more—maintain the right things better. The discipline is in making that distinction asset by asset and then following through.

In the next chapter, we'll move from *what* to do to *how* to embed it—integrating these choices into daily work so they stick and scale.

- Materials coordinators ensure the right parts and tools are on hand before work begins. They close that gap between good planning and scheduling to drive schedule success."

When any of these roles is missing, chaos creeps in—trades stand waiting, jobs stall, priorities shift. Training and sufficient CMMS access are vital for them to manage effectively.

Engineers and Technologists

Maintenance and reliability engineers play complementary roles. Maintenance engineers shorten downtime; reliability engineers lengthen time between failures.

- Maintenance engineers focus on keeping downtime short—solving immediate technical problems, finding substitute parts, refining procedures.
- Reliability engineers focus on making failures rarer and less painful—root cause analysis, proactive strategies, and design feedback.

Technologists often bridge the two worlds. Their broad technical background and comfort with systems make them excellent planners or reliability analysts. They ask, they learn, and they integrate.

For example, technologists make great leads for PMO and RCFA involving trades input.

Training: The Ultimate Nondiscretionary Cost

Training isn't optional. It's insurance. Untrained people working on high-value assets are accidents waiting to happen.

If turnover is high, fix that—don't use it as an excuse to skip training. Skills, knowledge, and mentorship are the foundation of safe, reliable performance. Without them, all the IT and AI systems in the world won't save you.

Here are some important considerations for you as a manager.

Training is nondiscretionary. New hires don't know what they don't know; managers often underrate maintenance because they've never seen top-quartile performance. Set higher expectations, show what "good" looks like, and fund the learning.

In some cases, where employee attrition has been high, companies are loath to spend on training because they know there is a good chance the employee will leave soon anyway. Their choice is to have untrained employees working on their valuable assets and highly prone to making mistakes, rather than accept the cost of training. Digging into the causes of that attrition might pay off more and remove the excuses for being neglectful. It's almost certain that your maintenance leadership cannot achieve this on its own. As an executive, you'll need to be involved.

Training and education are needed to forge an effective maintenance workforce. Because so many people arrive in maintenance with little understanding of what is required, they also arrive not knowing what they don't know.

When they are new, you can't rely on them to tell you what training or development they will need. You'll need to have some idea. As they gain experience, they'll see where they are lacking relative to your expectations, but if your expectations are low, you'll find that they don't learn much.

Business managers in finance, human resources (HR), and elsewhere also have low expectations of maintenance. They have had no education in maintenance and what it can achieve. Many see maintenance as an expense, a cost center to be minimized as a method to increase profitability. They probably trust and believe that you (the specialist) are trained and knowledgeable. They expect you to know what to do.

Too often, the uninformed are leading the uninformed—at least until some experience is gained. Even then, if none of them has seen high-performing organizations, they may not know what is achievable. Companies don't share much information among each other, nor even within operating locations in the same company. High performance in one operational site may be ignored by the others simply because they don't know the good site is any different or any better. Ego plays a role, too: as experience is gained, we tend to believe we "know it all," and that makes us less open to learning and new ideas.

Those are barriers that need to be broken if your maintainers are to produce high-performance results for your business.

Without skills, knowledge, and mentorship, no system will save you.

Competency Development

Not only is training non-discretionary, I recommend going one step further. You must develop competency, not just provide training. Many organizations formally define jobs and roles. Some include a definition of needed competency for them. But most stop short of operationalizing it. To do so, you need to give thought to:

- Role profiles (job descriptions) linked to business processes and supported by skill matrices,
- Hiring criteria,
- Learning pathways (formal plus on-the-job),
- Hands-on assessments (not only attendance),
- Certification/qualification for critical tasks and methods,
- Periodic re-qualification
- Governance (someone who tracks it, ensures it is kept up-to-date, and approves exceptions).

This turns "training spend" into an asset-protection mechanism.

In some constrained environments, access to skilled trades, formal training and education resources may be limited. I've seen companies set up their own training centers for trade skills, supervisory skills, and even ongoing upgrading of skills. They design their programs to accommodate and work with the available human resources and education levels found locally.

In other areas, formal education is good, but not technical, and may not serve the industrial nature of our work well. It is necessary in these cases, to build capability intentionally: define role-based skill matrices, create mentoring ladders, certify critical skills, and protect time for training. It may be necessary to bring outside help in to start, and then train the trainers. Doing this for all roles can be overwhelming, so focus first on the roles that unlock the most value (planners/schedulers, supervisors, lubrication techs, reliability engineers), and then scale it up.

Even in developed areas, you may be faced with education systems that are producing graduates whose knowledge and skills are not well aligned with your needs. Working with local educators is an option if they are open to it. It is always wise to verify any skills and knowledge that you may be expecting new hires to arrive with.

Training and Outside Help

It's pretty easy to find technical training on methods (like RCM) and practices (like precision maintenance). If you want your trades to know how to use a specific model of laser alignment tool, for instance, the manufacturers provide that training, and it is very good. There are certification programs in vibration analysis, lubrication, oil analysis, ultrasound, thermal imaging, and nondestructive testing. Training for the necessary skills is pretty easy to find, but management of it all is another story.

There are many good books on maintenance and its management. As an author of the bestseller *Uptime: Strategies for Excellence in Maintenance Management*,[1] I can tell you that even that status doesn't mean all have read it. Maintainers don't tend to be voracious readers, unless the content is relevant to whatever job they have at hand. Managers and supervisory or superintendent-level staff do have a need to keep on top of what's going on in their fields, but they often lack the time to do so. They are very stretched. Like most of us, no one wants to go to a doctor who isn't up on current diseases and treatments, and most certainly not if that doctor is supposed to be a specialist. Your maintainers are specialists, and it is not at all unreasonable to expect them to keep current in their field. Reading isn't a huge burden, and it can really open people's eyes, but waiting until they have a need to get abreast isn't a great strategy to sustain high-performance levels.

If reading won't work, you might try training courses and conferences. Conferences are great places for networking with others in our professions. For maintainers and reliability people, it may be the only place (or one of very few) where they can meet others in the same field, have discussions, compare notes, and learn from their experiences.

There are a few good training and certification programs in maintenance, its various components (like planning and scheduling), and reliability. There are no actual standards or established academic norms, so it's a bit challenging to find good training and education for your people in leadership and management roles.

There are consultants in maintenance management, and if you think you might have a need or opportunity to improve, you can call on them. Some are truly professional, certified and experienced; others are not. A number have worked in the field, but most haven't been in more than one or a few operating sites. Do they truly know what good looks like? One who just retired from your operation would certainly be helpful as a spare hand or to fill a gap, but can they really help you lead a change in performance?

To make those important transformations, look for practitioners with multi-site depth, evidence of sustained results, and the ability to transfer capability—not just fill a gap.

Cross-Functional Teamwork

Teamwork is often seen (and rightly so) as one of the best ways to get great results. The Toyota Production System, Total Productive Maintenance (TPM), and a host of lean maintenance proponents swear by it as the way to keep equipment running with minimal downtime. In most cases, a degree of management decision-making and execution of maintenance is in the hands of shop floor teams comprising operators, maintainers, and someone to help with parts and purchasing. The teams are often autonomous, meaning they make their own decisions.

In some cases, there is no overall maintenance department, manager, superintendent, or even budget. And it works! But not always. Unfortunately, nothing seems to be the one best way to use everywhere.

Teamwork Japanese Style

The autonomous team approach works very well for manufacturing industries, such as automotive, where there are many work stations or production lines. The teams have "ownership" of their area. Generally, those areas are fairly small in comparison with an entire plant, so there are several teams, one per area. The team members share responsibilities for all activities in the area: operations, troubleshooting, maintaining, repairs, and even improving through small projects. Although they have formal meetings and methods they use, they appear to work very well and informally as a team. You can't tell who is an operator and who is a maintainer. They also get along with each other and cover for each other if someone is off.

Management techniques are very visual, they will spend time observing to identify wasted movement, materials, and so on. The workplace, where things happen, is *gemba*. Observations are carried out there and improvements are made there. There's no one sitting around in an office deciding what should be done and how; it's all done at gemba.

Kaizen is how they make improvements—small, incremental improvements. They don't do big projects or training programs followed by analysis sessions. They are constantly on the watch for *muda,* or waste. The eight forms of waste they look for are defects, overproduction, waiting, nonutilized talent, transportation, inventory, motion, and extra-processing.

- **Defects.** Result in rework or scrapping of whatever is being made. Find out why and eliminate them.
- **Overproduction.** Refers to producing more than is needed by the next person or station in the manufacturing process; it builds up as work in process between stages.
- **Waiting.** Idle time in machines, people, or materials waiting to be used. Delays in waiting for inputs or for the next station to take what you've produced result in waiting.
- **Nonutilized talent.** Wasted human potential where employees' skills, talents, or creativity are not being used to full potential.
- **Transportation.** Any movement of materials or product that isn't absolutely necessary.
- **Inventory.** Excess raw materials, work in process, or finished goods tying up space and capital.
- **Motion.** Unnecessary movement of people, such as walking, reaching, or searching, which can cause fatigue or lower productivity.
- **Extra-processing.** Performing more work on a product than required by the customer, such as overspecifying or adding extra features.

Production lines operate at a fixed pace or *takt*, so that materials, inventory, and time are not wasted and do not allow for wasted motion, overproduction, or transportation.

When a fault occurs, each person in the team is empowered to stop the whole line while it is fixed. Most repairs and fixes, to product or the line itself, are small and quick when you stop for each one. Problems don't accumulate into big ones. Everyone is watching for them and can act.

This works incredibly well where breakdowns tend to be small and easily repaired, and where the whole team can be involved. It doesn't seem to work as well in places where the equipment is massive, is difficult to work on, and requires specialized skills. It also doesn't work well if the workforce isn't used to teamwork.

Cultural Differences

Teamwork is exceptionally successful in Japan, where there is a culture of collective harmony. It prioritizes group success over individual achievement. It uses a strong sense of loyalty, respects hierarchy, and practices shared responsibility. Success and failure are attributed to the team, not individuals. It fosters strong relationships, has everyone contribute, and requires careful planning and support. In other cultures, teamwork isn't as easily achieved nor as successful.

For instance, in Western cultures there is a strong emphasis on individual performance and success, not teamwork. We love our heroes. We have individual roles and responsibilities, and we tend to challenge authority. If we think we have a better way, we go ahead and do it. If it fails, we take blame on our own, not as a team. If it works, we want the recognition—again, individually, not as a team. We have winners and losers, the Japanese do not. We also tend to operate without planning, let alone the meticulous planning you'll find in Japanese culture. Companies have tended to be stingy about rewards, so continuous improvements seldom come from the workers, they come from engineering or management.

Attempts at lean, TPM, and the Toyota Production System methods have often failed, even with direct help from Japanese parent companies. Our Western culture does not appear ready yet for the full-blown teamwork approach that is so successful for Japanese companies. These aren't value judgments but they do explain why copy-and-paste lean often fails.

Teamwork Is Still Valuable

Despite those cultural differences and tendency for failure in such a broad cultural shift in the West, teamwork is needed and works well to a degree. My observation is that it works where there is a common "enemy" or problem to be solved, then it fades away and is rarely sustained. Indeed, for success in maintenance and reliability, you'll need teamwork among departments: maintenance, engineering, operations, stores and inventory management, and purchasing. Collaboration won't be natural though, so shared performance indicators and rewards will be needed to encourage it.

Our siloed organizational structures may work well for financial budgeting purposes, but they encourage silo thinking. Like an assembly line, each station or department has a specific role and in theory, if everyone does what they are supposed to do, it all works well.

But in reality, everything that needs to happen is not quite so well defined (no meticulous planning), and managers have responsibility for a broad function that has somewhat fuzzy boundaries but a strict budget. They must comply with budget, and it's all about spending. Their only impact on profit is to lower spending. Accountants might like that, but in finance they know that revenue opportunities can be easily missed. Managers have no revenue responsibility, and so an improvement that will increase revenue overall gains little support from managers who are worried only about their budgets.

Structures reinforce behavior: Silo budgets optimize local cost, not total value.

There's a tendency to stall or ignore anything that isn't in the budget, even if it will produce a savings bigger than the spend or increase revenues. You need to get to the level of an executive or a general manager who has profit-and-loss responsibility (business performance) before such improvements get a lot of attention. Silos get in the way. If effort has to happen in one department (say, inventory management and stores), but the savings accrue in maintenance, there is nothing to encourage the needed collaboration to make it happen. Indeed, budget compliance and any penalties for noncompliance discourage such collaboration.

While teamwork has the potential to drive improvements and results, Western culture and business structure, with their silos, management blinkers, and budget focus, tend to undermine anything that requires teamwork. The way around it is to confront it head on and find ways to encourage that teamwork. That is unlikely to happen so long as we are stuck in those management paradigms and managers remain both unable and lacking in authority to change it. Making that happen requires direct intervention and support from the executive levels. Any broad improvement initiative that doesn't have the executive level sponsorship and active support is not going to get far.

That is not meant to be discouraging—quite the opposite—it shows that it is possible, but it requires an approach that goes beyond that possible by the managers alone. At best, they can change and improve only those areas for which they have direct responsibility, and then only if they have authority to move budget around to do it.

Managers can collaborate, but they are often contrained by their budgets, and the metrics they are measured by. If an improvement will result in mutual benefit, then it can go ahead. For instance, RCM requires participation of maintainers and operators. It will result in lower maintenance spending, fewer disruptions to operations, and hence increased ability to produce. Once those two groups realize that it is good for them both, they can collaborate on the provision of personnel needed for the training and analysis work and the time to do it, then later achieve the benefits together. I've taught many RCM courses and led many analyses. The operators are usually the ones who see the greatest benefit from this process that was originally seen as a maintenance-only initiative. In most cases, they are the first to name the next systems on which it should be applied.

Shared key performance indicators (KPIs) like availability, schedule compliance, rework rate, used in joint planning rituals, and executive sponsorship turn cross-functional intent into everyday practice.

If you believe you can, then you can!

Final Thought

Reliability lives and dies with people: Equip them, and they'll deliver the asset uptime you want. Ignore it, and you'll pay the price—every time.

CHAPTER 7

PERFORMANCE MEASUREMENT AND BENCHMARKING

Measurement gives us insight, but it can also mislead. The most useful metrics in maintenance and reliability don't just count what happened—they shape how we think, plan, and act. As W. Edwards Deming reminded us, not everything that matters can be measured, and not everything that's measured matters.

This chapter explores how to choose the few metrics that truly drive performance, how to link them to the behaviors that create value, and how to see measurement as a tool for learning rather than judgment.

W. Edwards Deming said, "It is wrong to suppose that if you can't measure it, you can't manage it—a costly myth." He is telling us that you don't always need to measure something to improve it. However, many only refer to a subset of that quote, "If you can't measure it, you can't manage it." That leads in exactly the opposite direction!

Having been a manager, I agree with Deming and learned long ago that you really don't need to measure everything. For instance, if you want people to feel appreciated and motivated, you won't be able to measure your results. If you publicly acknowledge a job well done or other achievements, they will feel good about it, and by watching behavior you can see changes in performance—without a metric. Sometimes I get a good laugh when I see measures being used as indicators of something that can't be measured—like employee engagement. You can see it, sense it, and feel it, but good luck quantifying it!

As we've seen elsewhere in *Steadfast*, performance is as much about human behavior as data. In this chapter, we will focus on what you can measure, but keep in mind that you don't need to measure everything to make improvements. In fact, measuring less can get you what you want sooner. People don't really like being measured, because it means they'll be compared and judged. Measures that impact people directly can create a winner-and-loser dynamic that can actually be counterproductive. Whenever you consider a metric, also consider how it will be received by those who work for you.

Less Is More

In maintenance and reliability, there are well over a hundred possible metrics you could be looking at. Terry Wireman[1] even published a book on just this subject. The trick is to use a few, not many, that meaningfully change behavior to achieve the results you want.

Back in the Vietnam War era, F4 Phantom fighter pilots used to get distracted by too many gauges. They were originally installed for the backseater in the cockpit who was responsible for the "fighting" aspects of the jet. The F8 Crusader also had a problem with pilots becoming too distracted by the many gauges in front of them whenever they were engaged in a dogfight. In both cases, nonessential gauges were taped over. In maintenance too many metrics can distract from what really matters.

In cars, we used to have just a few indicators, mostly analog (e.g., speed, fuel) or warning lights (e.g., engine hot, oil warning). Today, we have largely digital displays and a whole lot more to look at. Some of it comes on only when needed (e.g., back-up camera), but many are always on. When I'm driving, most of them get ignored. If possible, I scroll to menu items with pictures or less data to avoid being distracted. Do I really need a list of upcoming exits on a global positioning system map screen when I already know my way around? It's too bad those digital displays don't have an off button.

Input—Process—Output

If we view maintenance and reliability as a sort of "process," we can depict it as a box that has inputs and outputs. We'll get into each in this section, but in general, inputs are what we need to feed the process: essentially money in the form of parts, materials, and labor (contracted or otherwise). Outputs are what we want to achieve: availability of the assets (enabling production), asset uptime, or a certain level of cost per unit produced (a combination of output and input that indicates the efficiency of my spending). In the middle we have the process that we control.

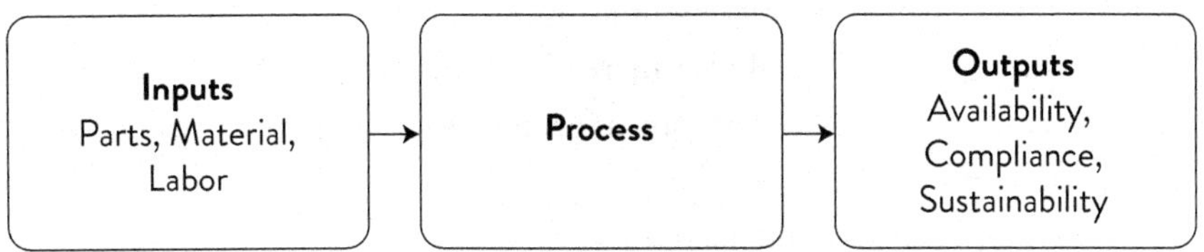

FIGURE 7-1 Input—Process—Output Model

Inputs

Costs always attract a lot of attention, especially by accountants, and anyone with a budget, but in truth we don't actually control them. They are a result of what work we had to do to achieve the output that was desired. If you have a breakdown and operations is pushing for a quick repair, you'll spend whatever you need to spend to make that happen. Costs only go down if we work "smart" or efficiently, changing what we do in the process.

To lower costs, we need to lower the amount of work we do: materials, parts, contractors, overtime, rentals, and so on. If you lower any arbitrarily, however, you'll impact negatively on what you can do in the process and sour the outputs.

We can best achieve lower costs by being more efficient, and reducing the amount of work we do by focusing only on the right maintenance.

Outputs

In operations, the big interest is the output: availability or asset uptime. As maintainers, we have a duty to deliver that availability so that operations can use it. We sustain productive capacity. Operations may want higher availability than you are currently achieving, but it is not something you can just dial up or down at will. Like costs, it is a result that is dependent on what work you do and how well you do it.

By working efficiently, we can lower costs, and we can shorten repair times. Better planning, scheduling, and schedule compliance will reduce maintenance time and resource use.

We can also reduce the amount of work by focusing on doing the right maintenance. Shifting to proactive methods like condition monitoring requires less downtime, involves fewer equipment interruptions, and will result in fewer breakdowns. By doing that, we will also increase availability and hence productive capacity of the assets. These relationships form the basis for a simple value map of maintenance performance.

A value map[2] helps us understand where we need to focus our attention (see Figure 7-2).

Lowering costs and increasing output requires us to focus on the work we are doing (effectiveness) and how efficiently we do it. As we do less work and more efficient planning, we reduce materials consumption and the spares we need to keep on hand.

Process

If you remember the circles of influence, we control a few things, have influence over others, and have interest in much more. With maintenance and reliability metrics, we are interested in what our operation produces. We influence that by delivering availability. We are interested in profitability; we influence by lowering our costs. We do both by being more efficient (doing maintenance the right way) and more effective (doing the right maintenance).

It is the maintenance processes we employ—and how well we execute them—that drive efficiency and effectiveness.

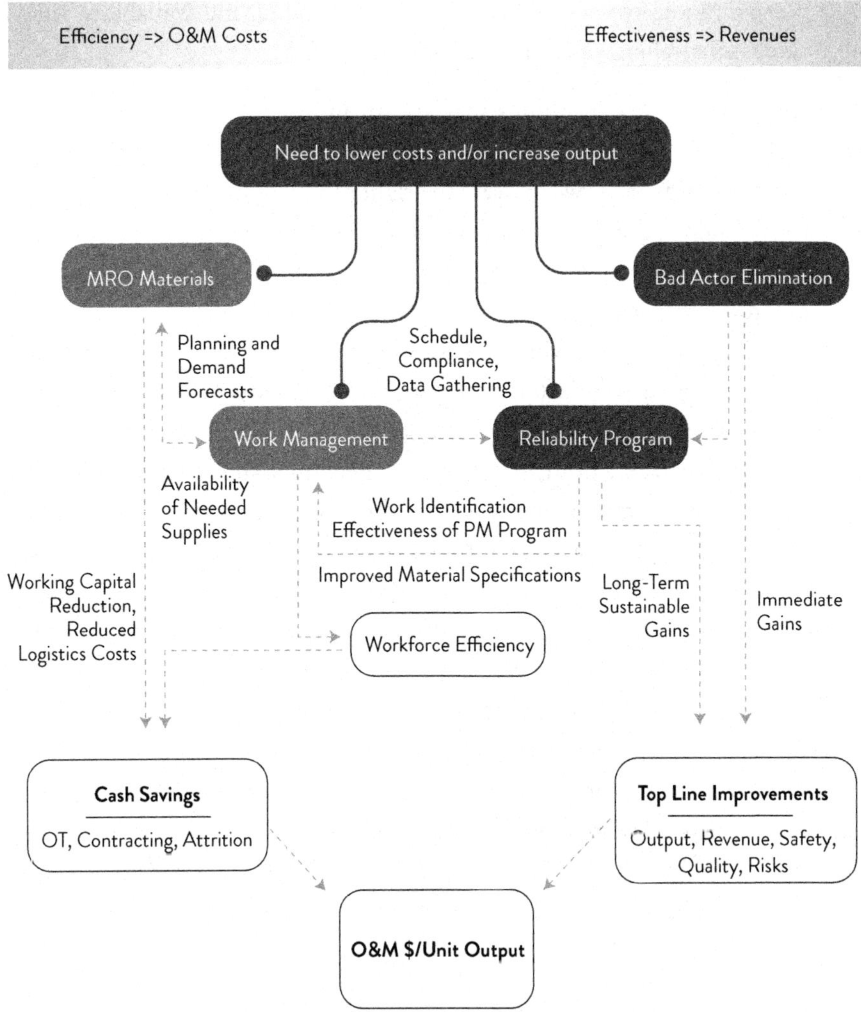

FIGURE 7-2 Maintenance and Reliability Value Map

Leading and Lagging Indicators

Some metrics indicate what we can expect in the future; some show what we have experienced.

- *Leading* indicators tell us what to expect.

- *Lagging* indicators are "after the fact," like spending and availability. They are results, telling us what we achieved.

Leading indicators are things that give us insight into future results. For example, if we are planning a greater portion of our work, we can expect the cost of that work to go down. If we increase the right proactive work, we can expect to see fewer breakdowns and less spending on repairs. Those changes may not happen immediately, but they will follow sometime after we adjust the levels of planning and proactive work.

I find that availability (which measures uptime relative to total time) will tend to change fairly soon after you make improvements that reduce breakdowns. It shows up in performance: The asset just goes on longer, operators realize quickly they can rely on it, and you don't need to wait for a breakdown to measure it. Some like to measure reliability directly, but it is actually trickier to measure, and it changes more slowly. Imagine an item with two years mean time between failures (MTBF). If you improve it by even a month, you'll need to wait for a few years to see the change—a big lag in that indicator. That's why leading indicators are more useful for managing change in real time.

There are also metrics that indicate the capability to attain high performance. Such "capability metrics" act as leading indicators; e.g., percentage of critical roles filled with qualified personnel, re-qualification compliance for safety/precision tasks or roles, planned work accuracy linked to planner competence, and sampling compliance linked to oil analysis program maturity. These and other similar metrics can predict reliability outcomes far earlier than many others.

Metrics and Human Response

Ultimately, our performance metrics are there to help us decide, act, and provide feedback on changes that are intended to guide our improvement efforts. Those efforts cannot be just one-sided. For instance, metrics intended to guide only improvements in shareholder value will fall flat: only shareholders and boards they hire really care. They are not the ones undergoing the change. The question of "What's in it for me?" must be answered for everyone who will participate in the change.

The answer must provide some benefit that exceeds the perception of pain or risk that will be tolerated while and after making the change. If the change puts your job at risk, you are not likely to be supportive. If it saves money for the company at the personal expense of less overtime pay, you won't be supportive.

We'll talk more about change later. Bear in mind that when we measure, we invariably judge and make decisions. Those are likely to be taken with some measure of suspicion by those being measured. *Be careful what you measure!*

There are people who take great pride in what they do, how they do it, and the results they generate. If they can see a "better way," they are likely to get onboard soon. Others however, are primarily motivated by the paycheck. For them, inefficiency means more hours, more overtime at higher pay rates, and more money for them to spend on whatever they love and truly motivates them—family, sports, and so on. Your workforce has both.

In practical terms, these behavior changes look like this. For changes in costs and availability, we need people working more efficiently and doing the right work. Planners will need to plan, not chase parts. Supervisors will need to stick with schedules. Engineers will need to do analyses to determine the right maintenance to do. Your planner, scheduler, or parts coordinator will need to confirm availability of parts, and so on, for planned jobs on the schedule for the following week. Trades will need to provide feedback to planners about the accuracy of the plans. Planning and inventory management will need to work together to determine accurate demand rates for needed parts. Trades, supervisors, and planners will all need to stop hording parts in their own stashes.

Recognize that for the performance you want, you'll need new behaviors. Those behaviors will eventually become habits, but until they do, you may find that people resist the changes. It may be necessary to tell and show them how the new behaviors will benefit them.

For instance, there will almost certainly be a lot of resistance to "no more parts stashes." You'll need to have stores and inventory management help you explain to them how those stashes are not helping and demonstrate that you are doing something new so that the stashes are no longer needed. Show them how the change makes their work smoother and reduces frustration.

Any metrics you choose to monitor should be process oriented and within the control of those in the processes. Individuals in your department who work in those processes need to understand how what they do impacts those metrics.

For instance, planners need to recognize that by planning more jobs, the percentage of planned work overall can increase. They should focus on that and not on the day-to-day chasing of parts. While those parts are important to a job being done today, they take the planner away from his future-focus, and do nothing to improve overall performance of the department.

Supervisors will need to accept that schedule compliance is important and really matters if the proactive work is to achieve better availability. Supervisors should resist real-time reactions to operator requests and reinforce proper work-management channels.

Achieving any sort of substantive change won't occur by merely measuring. Measure the wrong things (at least wrong in the perception of your workforce), and you could drive behaviors you don't want. If you dig into what leads to labor disputes, you can often find metrics that are good for the company and bad for the employees. Labor disputes often trace back to metrics that reward the company but punish employees. There's a good reason for the old adage that companies get the unions they deserve.

Some Key Metrics

Here are some of the metrics I find most useful to monitor—and what I watch for in each.

Inputs (and Inputs Related to Outputs)

Inputs are what you spend on: materials and labor. Rental equipment may be in the mix as well, but it's ultimately a material spend (expense, or depreciation of what you purchased). On their own, you can't control costs, but they can give lagging indication of how well you have done. Trends can show directionally whether you are improving or regressing. As a rule, maintenance spend on its own is meaningless. You will spend whatever you need to keep things running; spend proactively with some discretionary "investment" in performance, and you'll spend less overall.

Common examples of metrics include:

Maintenance Cost as a Percentage of Your Operating Budget

- If operations activity and production increase, asset utilization increases along with wear and tear. You can expect an increase in maintenance roughly in proportion to your other costs. Trends in this can show deteriorating or improving performance and lead you to investigate further.

Maintenance Cost per Unit Produced

- The selling price of whatever you produce (ounces of gold, ore, vehicles, or consumer goods, etc.) can change. Your maintenance costs may vary but on a per unit basis stay more or less the same.
  - If they trend up or down, you are doing something wrong or right, and those trends can point you in the direction of what to watch for.

Percentage of Maintenance Spent on Materials and Labor

- Maintenance consumes labor and materials. It may be spent through contractors or your employees, on parts and materials through contractors. In my experience, they will be roughly proportional to each other.
  - Here in North America, I find that those two costs are roughly in balance: half and half. Elsewhere labor costs might be higher (Europe) or lower (Asia), and the proportions change, but the ratio likely stays the same.

Outputs

Some outputs are considered with inputs to give a sense of how "cost-effective" your spending has been, as shown earlier. As with inputs, you don't control outputs directly, but you do influence them with the actions you take in the processes that you do control. Other outputs that you might find useful are:

Availability

Here's how availability can be broken down in practice.

- It is the percentage of uptime relative to total time. It can be measured several ways. If you subtract all the time the asset is in your care (repair times, active maintenance working time), you get "mechanical availability." If you consider the time the asset is there for operations—subtract the travel time to and from the shop, all the downtime from when the asset is no longer in use until it is available for use—then you get "operational availability." You probably don't have all the data for this in your CMMS, but you should have some of it.

  - Figure 7-3 shows a timeline (total time) and various parts of it that are all measurable. You probably have mean downtime (MDT) and mean time to repair (MTTR) readily available as a result of recording maintenance times in your CMMS.
  - Your operations people probably have a better idea of total downtime because they really care about it. They will measure downtime starting from when equipment breaks down until they have it back and ready for use.
  - Note that "in use" is a subset of total uptime. Your car in the parking lot is available, but only in use when you are driving it. Plant equipment is similar. It can be available and ready to use, but not in use. Utilization is a measure of how much it is being used.

- As a rule, availability is easy to measure. It is simply Av = uptime / total time.
  - I prefer to use total time (e.g., all 8,760 hours in a year) and not "scheduled" production time. Scheduled time can be controlled, and some like to "game" the numbers. They can't fake calendar time.
- You can also simply measure the asset's uptime, but make sure everyone (maintainers and operators) knows exactly how you are measuring it.

Reliability

- Some like to measure MTBF for equipment, but it can mislead; you need to know that the intervals are actually between failures, not just maintenance actions that may or may not be related to failures. If your maintenance records are lacking, you will struggle to measure this.
- MTBF is also valid for "failure modes," and your maintenance records deal with only equipment-level data.
- MTBF can also have a long lag time before you see changes. Unless you have a large number of like equipment and combine their data (e.g., a

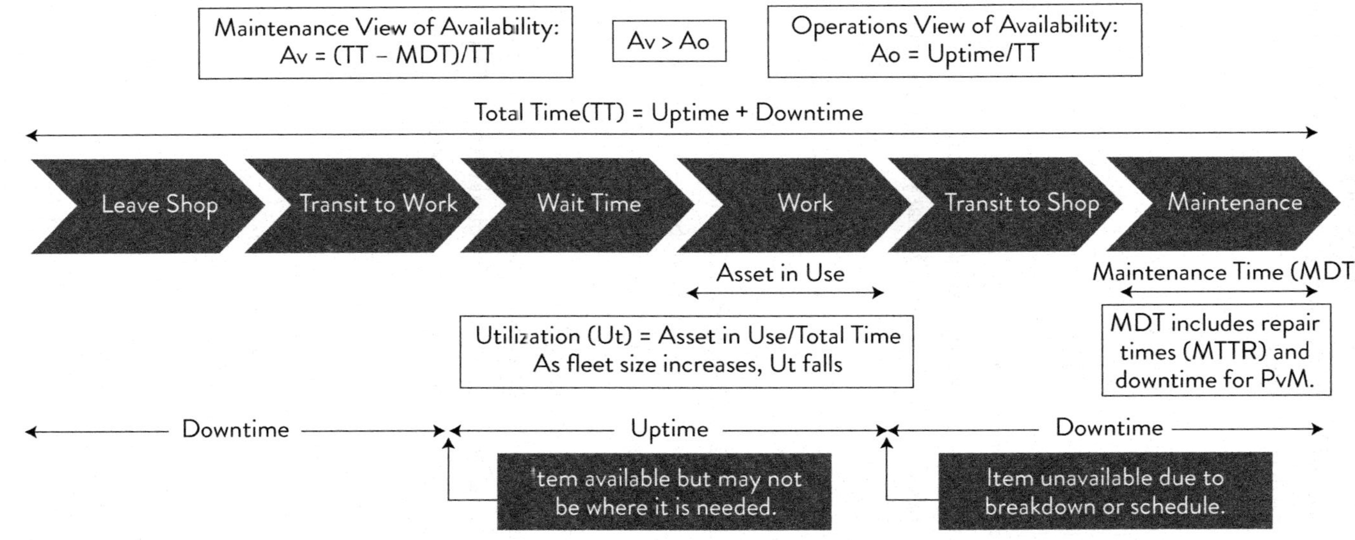

FIGURE 7-3 Availability Explained

fleet of the same asset such as variable frequency drives (VFDs)[3] in a plant with many variable-speed conveyors, such as a postal or parcel handling facility), it will be slow to change.

Production

- Output is ultimately what you want, but it can be variable due to conditions other than equipment availability. For instance, you may have a slow market for your products and production is cut back due to low sales. Or you may be able to meet market demand in only part of the week (e.g., five days instead of seven). If you have breakdowns and use one of those weekend days to catch up, you won't really see much of a trend by measuring output. It works best if you can sell all that you can make and operate around the clock.

Process Metrics

These measure processes, activities, or their direct outputs that you can control. You cause these to go up or down very directly. Some depend on coordination with other managers, but together you can control them. Teamwork is important to those.

Refer to Table 7-1 for benchmarks. A number of process metrics there give a good idea of what works well. Note that top-quartile numbers are not necessarily the best that are achievable. In my early petrochemical work, we had availabilities in our processing plants that consistently exceeded 99 percent (well above what the table below shows).

Across industries, top-quartile performers share the following traits. They:

- Are safer (have one-fifth the accidents)
- Carry less parts inventory (about 23% less)
- Have half or less the maintenance costs relative to the value of operational assets
- Achieve higher availabilities (average 17% better)
- Have greater overall equipment effectiveness (63% greater)
- Enjoy greater wrench time from a more productive workforce (68% better)

- Do much more planned work
- Have greater schedule compliance
- Do more of their work with formal work orders
- Use operators to do some of the maintenance work (usually the simpler tasks)
- Have fewer mechanics (41% fewer)
- Get far more improvement ideas from their workers
- Have higher stores turnover and much better service levels
- Get more of their materials out of stores than by direct purchase
- Have fewer supervisors, planners, and engineers
- Spend less on training than those who have a lot of improvement yet to achieve

These metrics and the conclusions agree with my own experience in the field with many companies.

Here's a caution: Don't expect to improve in only one area at a time. The various activities to achieve them are very much *interdependent*—they work together, not independently. The improvements you need to make to achieve those top-quartile metrics (or better) must be made across the board.

For instance, to achieve high levels of availability, you'll need more proactive work (but not too much), more planned work, and high levels of schedule compliance. In turn, you'll need fewer breakdowns disrupting the work. PMs must be effective and service levels from stores, high. Those require excellent plans and good communication among planning, inventory management, and purchasing. The goal is to have what you need to sustain high availability, not what is cheapest.

Benchmarking

Many people want to achieve "benchmark" performance, but benchmarks are little more than ranges of metrics gathered across multiple organizations. They really want to achieve top performance, but top performance for your organization may mean something very different in another. Remember, the only benchmark that truly matters is your own improvement.

Benchmark studies can be difficult to find. The simple truth is that very few are published, because they've mostly been paid for by companies who are looking for comparisons and then treated as corporate secrets. I've participated in several, and the biggest problems are *different definitions and poor data.*

In truth, the only comparison that really matters is your own change in performance. Wherever you are today, regardless of whether it is a top performer or a reactive low performer, what matters is how much you improve. Look for metrics that matter to your business, come up with definitions that you can actually measure against, baseline where you are, and measure the trends thereafter.

Different Definitions

Companies tend to measure things differently, making their questionnaire responses nearly impossible to compare. Even when providing detailed definitions, the results coming back don't comply. In some cases, they informed us of how they determined their versions of the metrics so we could adjust them for comparison, but often that wasn't the case.

To address the different definitions, some efforts have been made to standardize. The best known is probably the US-based Society of Maintenance and Reliability Professionals (SMRP).[4] They include standardized definitions of how to measure them. They are metrics that can be used, and a few include targets. Here is a sample of those:

Work management:

- Wrench time: 55 to 65 percent
- Schedule compliance: 80 to 90 percent
- PM compliance: more than 90 percent

Some cost metrics include:

- Maintenance cost as a percent of replacement asset value (RAV)
- Maintenance unit cost

Inventory metrics:

- Stock maintenance, repair, and overhaul (MRO) inventory value as a percent of RAV
- Stores turnover
- Stockouts

Reliability metrics include:

- Overall equipment effectiveness (OEE)
- Availability
- MTBF (discussed earlier)
- Mean time to repair (MTTR)
- Mean downtime (MDT)
- Mean time between maintenance (MTBM), includes all maintenance interventions
- Proactive work percentage (preventive, predictive, and proactive repairs resultant from condition monitoring aimed at avoiding failures and their consequences)

They also speak to organizational structure, skills, and qualifications:

- Training hours per employee
- Rework (an indicator of how well the employees actually do with their work)

Data Quality

In maintenance, data quality is usually quite poor. It has long relied on inputs from the field on work orders: often poorly written, minimalistic data, little if any text description, poor practices in supervision and collection of the data, and delays in data entry, making any time stamps irrelevant. Barcodes, quick response (QR) codes, and mobile devices are contributing to improvements here, but the biggest problem is with those who collect the data: field technicians who would rather be "working" (on the tools) than recording data on forms, especially when they rarely see that data used for making improvements. Without accurate data, even the best KPIs become unreliable.

Aside from SMRP, a few other organizations and publications have published studies with metrics. The most comprehensive I've found was published in *Iron and Steel Engineer*[5] in 1998. Table 7-1 depicts that study. The reader is referred to that study for its definitions. If you cannot find a copy, then reach out to me. While the data are historic, the patterns remain instructive. Table 7-1 summarizes the metrics and ranges of performance.

TABLE 7-1 Maintenance Benchmarks (dollars are adjusted to 2025)

Benchmark	Bottom	Third	Second	Top
	Results			
OSHA Injuries per 200,000 Hours	>5.5	5.5–3.1	3.0–1.0	<1.0
Stores as % of Replacement Value	>1.3	1.3–0.8	0.7–0.3	<0.3
Maintenance Cost as % of Replacement Value	>5.0	5.0–3.2	3.2–2.0	<2.0
Discrete	>5.0	5.0–3.2	3.2–2.0	<2.0
Batch Process	>3.5	3.5–3.0	3.0–2.5	<2.5
Chemical, Refining Power	>3.5	3.5–3.0	3.0–2.5	<2.5
Paper	>4.9	4.9–4.5	4.5–3.5	<3.2
Availability	<78	78–84	84–91	>91
Discrete	<78	78–84	84–91	>91
Batch Process	<72	72–78	78–91	>91
Chemical, Refining Power	<85	85–89	89–94	>95
Paper	<83	83–85	85–89	>94
Overall Equipment Effectiveness	N/A	48–48	48–78	>78
	Process			
Mechanic Wrench Time	<31	31–41	42–52	>52
Percentage Planned Work	<65	65–78	79–90	>90
Request Compliance Percentage (ability to meet overall schedule)	<88	88–95	96–99	>99
Schedule Compliance Percentage (daily compliance with schedule)	<68	68–77	78–90	>90
Work Order Discipline Percentage (hours covered by Work Orders)	<54	54–59	60–84	>84
PM Percentage by Operations (indicative only)	<15	15–25	25–35	>35
Replacement Value ($million) per Mechanic	<6.4	6.4–7	7–9	>9
Suggestions per Mechanic per Year	N/A	0–0.4	0.5–4.0	>4
Stores Turnover	<0.5	0.5–0.7	0.7–1.4	>1.4
Stores Service Level (% of requests filled when requested)	<85	85–89	90–94	>95
Contractor Cost Percentage (indicative only)	<8	8–13	20–30	>40
Stores Issues/Total Material Percentage	<82	82–86	87–90	>90

Benchmark	Bottom	Third	Second	Top
Training and Staff				
Span of Control (team environments can distort)	<9	9–17	18–40	>40
Mechanics per Effective Planner	<25	25–59	60–80	>80
Replacement Value ($million) per Maintenance and Reliability Engineer	<50	50–120	120–140	>160
Mechanics per Plant Worker (can vary with level of automation)	<32	32–21	21–10	>10
Total Craft Designations	>7	7–8	8–9	<6
Training Hours per Mechanic	>80	80–70	69–40	<40
Training Cost per Mechanic	<6000	6,000–3,600	3,600–1,000	<1,000

The study was created with data from some 140 high-performing companies (financially). The ranges of performance were divided into quartiles. *Maintenance performance at the low end (bottom quartile) wasn't necessarily all that good,* yet the company still achieved top-quartile financial performance. Bear in mind that although maintenance and reliability contribute to overall performance, other factors (e.g., commodity prices, overall market conditions) may have greater influence on financial results. At the time of the study (1998), global financial markets were in a downturn.

One metric (maintenance as a percentage of sales) has been removed. Sales dollars varies with product pricing and is not related to maintenance cost.

Dollar amounts were from 1998 and have been converted here to 2025 dollars. Participating companies were from a variety of industries, some using lean manufacturing methods where conventional maintenance and operations divisions are blurred. That impacts a few metrics—particularly those with very wide ranges. Definitions for most of the metrics are well known. For a few, I've added explanations in brackets. Refer to the original article for details.

In the previous section, I've interpreted the process metrics for the reader and provided a caution on using just a few of the numbers. If you set your department's performance targets at too much of a "stretch," you can create a sense of "We'll never get there" on the part of your departmental staff. Slow and steady increments will get you farther in the long run than attempting to make massive leaps.

Keep it achievable for your people. Keep your expectations realistic. Give up on quarter-to-quarter thinking and short interval controls for such a massive undertaking. Improving from a level that compares to the lower quartile up to a level more like the top can take two to three years.

Performance measurement isn't about chasing numbers; it's about cultivating behaviors and processes that make those numbers meaningful.

The real power of performance measurement lies not in the numbers themselves but in what they inspire us to do differently. Metrics are only meaningful when they reflect process, behavior, and purpose—the things we can actually control and improve.

Benchmarking can show us what excellence looks like, but our true benchmark is progress: doing better tomorrow than we did today. Measure less, understand more, and let every metric serve a purpose that people believe in.

CHAPTER 8

MAINTENANCE, REPAIR, AND OVERHAUL SUPPLIES MANAGEMENT

The crew's ready, the job's planned, and the bearing's missing again . . . an all too familiar and avoidable situation. Maintenance can't succeed in isolation. Even the best planning, scheduling, and execution fall apart if the right parts and materials aren't available when they're needed. Too often, maintenance and materials management operate as separate worlds—one focused on reliability, the other on cost control. Yet both are working toward the same outcome: keeping the plant running safely, efficiently, and without unnecessary expense.

This chapter brings those worlds together. It explores how managing maintenance, repair, and overhaul (MRO) supplies—the bearings, belts, fuses, bolts, and tools that make maintenance possible—can strategically support reliability, reduce downtime, and control cost. When the supply chain and maintenance teams work as partners instead of silos, everything runs more smoothly.

Maintenance is only as good as the materials that support it. You can have a well-trained team, excellent procedures, and a solid plan—but if the right part isn't available when it's needed, everything stops. Every maintainer knows that feeling: The job's ready to go, the crew's waiting, and someone's scrambling to find the one bearing or seal that's holding it all up. That's why management of MRO supplies, although often owned by procurement or stores, is absolutely central to maintenance performance.

The Strategic Role of MRO Materials

Many organizations treat MRO inventory as an afterthought. It is used to support a support function (maintenance), and it isn't involved in making product. Accounting and most managers see it as an overhead. Yet it's one of the biggest enablers—or disablers—of reliability. Stockouts cause downtime, excess stock ties up capital, and poor-quality parts cause repeat failures.

To serve the business, we want complete alignment of the MRO supplies function with the maintenance function to sustain operations. This means viewing the storeroom not as a warehouse, but as a critical service provider to maintenance. The goal isn't to minimize inventory—it's to maximize producing assets' uptime at the lowest practical total cost. That's a very different and much more strategic mindset.

For instance, some organize parts in a storeroom by commodity, while others organize by equipment type—the latter being aligned with how maintainers think and look for their needed parts.

Managing the Supplies

Classification and Criticality

Not all parts are created equal. Some are cheap and easy to replace; others are expensive, have long lead times, or can stop production if they fail.

A good MRO system classifies items according to their criticality—how essential they are to safety, production, or reliability. As shown in Table 8-1, typical classifications might include:

- **Insurance items.** These parts or assemblies are held as part of your spare parts inventory and you would not typically expect to use them in the normal life of the plant and equipment, *but* if they're not available when needed, it would result in significant losses. These are often engineered parts that safeguard you against extremely long lead times or the original equipment manufacturer (OEM) going out of business.
- **A-items.** These are critical. They must be in stock or you must have a backup plan to obtain them quickly when needed.

- **B-items.** These are important but not critical: manage with lead-time awareness. Extended stockouts can lead to significant problems for production.
- **C-items.** These are typically low cost, and their absence has low impact. They are managed for efficiency.

TABLE 8-1 Selecting Material Classifications

	Lead Times ⟶	Very Long	Long	Medium	Low
Criticality	High	Insurance	A+	B	C
	Medium	A+	A	B	C
	Low	A	B	C	C

A+: Treat as insurance items if supplier could go out of business or if lead times could grow excessively.

Evaluating risk adds another dimension. If a part's absence can cause significant downtime, injury, or environmental impact, it is highly critical—even if it's inexpensive.

Mature organizations are using reliability-centered maintenance (RCM) to help in identifying the most critical failures, creating plans proactively, and using them to identify needed parts and how they should be managed.

Also, think in terms of repairable versus consumable items, and watch for obsolescence. Aging assets often have parts that suppliers stop producing, making proactive sourcing or refurbishment strategies essential. Some items containing elastomers or radiation sources may also have a shelf life.

Parts Planning and Forecasting

The best way to avoid "firefighting" for parts is to tie materials planning directly to your maintenance plans. Every proactive maintenance (PM) task, shutdown, and condition-based work order should include identified and reserved materials. The plans[1] for each job contain that information and hence the importance of having detailed plans. Those plans should indicate how much of each item is needed each time the job is performed, but they won't say "how often." That is in the schedule.

For preventive tasks, demand follows the schedule; for repairs, demand is unpredictable—but patterns emerge over time.

Maintainers can be their own worst enemy when it comes to having the right quantities on hand in stores. Inventory management looks for those patterns based on usage. If parts usage includes "direct purchase" items and items that were stashed away in a maintainer's hidden stores, that usage as seen by your stores and inventory management is distorted. What inventory management sees is far less than actual; stocked quantities will therefore be lower than needed and stockouts frequent. In one large integrated steel mill, our consulting team identified hidden stores that were roughly equal in dollar value (tens of millions of US dollars) to those "on the books" in the storerooms.

Using an AI-enabled analytic tool, we've consistently seen large-dollar opportunities to lower those hidden stores. By correlating what is reported as used on work orders with actual stores issues and purchases, what is and is not needed, and what's likely hidden in stashes are relatively easy to spot. Reductions in needed stores of up to 30 percent are not uncommon.

Those gains are impressive, but only one time—sustaining them requires more attention to processes and peoples' behaviors.

Forecasting doesn't need to be perfect—it just needs to be informed. Use:

- Historical usage data from the computerized maintenance management system (CMMS) and from your parts stores
- Upcoming planned PM work
- Failure rate trends from reliability analysis and historical usage for repairs

Predictive technologies can be a big help if you act on their findings in a timely manner. For example, if you're monitoring vibration and see a bearing trending toward failure, having that part ready is the difference between a planned repair and a breakdown. If the item is an A-item, you want it in stores; if it is a C-item, you can probably order it when the need arises.

Predictive insight is leveraged beyond reduction of downtime, if it also drives material readiness.

Inventory Management and Control

There's always a tension between minimizing inventory and ensuring availability. The trick is to balance service level and carrying cost. *The optimal balance depends on risk tolerance and production criticality.*

Set clear stocking policies:

- **What's stocked locally versus ordered on demand.** If you are in a remote location, you'll need more in stock than an operation in or near a large industrial center.
- **Consignment or vendor-managed inventory (VMI) arrangements.** Great for those C-items and lower-cost B-items.
- **Reorder points and safety stock levels.** Recognize that these can vary with lead times and as demand fluctuates when production levels change or when new equipment is added. Do not rely on maintainers to guess at the needed "min" and "max" levels—calculate them properly.

Shelf-life management is another silent killer. Lubricants, seals, and electrical components can deteriorate in storage. If your storeroom isn't climate-controlled, even more so. Poor storage conditions can quietly undermine reliability.

Even spares need care. Some items may require maintenance while in storage. Electric motors should be turned manually to avoid bearing damage from sitting in one position too long. Small prelubricated gearboxes should be turned as well. Some electronics may even need to be powered up in storage.

Procurement and Supply Chain Coordination

Procurement's job is usually to minimize cost. Maintenance's job is to minimize downtime. Those objectives often collide. Too often, the two functions don't understand each other's constraints, so they work in silos driven by conflicting KPIs.

Buying the lowest-cost part can lead to equipment breakdown once it is put into service. Bargains are seldom really bargains when it comes to parts. Cheaper is often lower quality and can be ill-suited to the intended purpose of the item. Saving pennies or a few dollars can be very expensive if it results in downtime on a production line or a missed customer delivery. Purchasing rules are often inflexible and can lead to some poor choices being made—when it comes to MRO supplies, penny-wise is almost always pound-foolish. When maintainers say "buy this one" they mean it. When buyers procure "or equal" it usually ends badly in terms of premature failures, poor fit, additional (now rushed) buying, higher ultimate costs, and lower plant availability for production (revenue).

Most maintainers have experienced a stockout when the job was urgently needed. Consequently, they learn to distrust stores and even purchasing. They take matters in their own hands by stashing parts away—off the books, and often out of mind, soon forgotten. Direct purchases against work orders will result in consumption that bypasses inventory management, understates actual demand, and undermines the efforts of inventory managers to have what is needed; that guarantees future stockouts. Maintainers seldom understand this and end up being the authors of their own misfortune. *Hidden stock undermines both trust and data integrity.*

The solution isn't to wrest control—it's to collaborate. A little bit of mutual understanding, perhaps even a little bit of cross-education, can go a long way. Maintenance can help procurement understand the true cost of poor supplier reliability or long lead times. Conversely, procurement can help standardize parts, leverage supplier relationships, and manage contracts that include performance guarantees.

Strategic sourcing is powerful when done right. Fewer suppliers, standardized components, and long-term agreements can all simplify operations and reduce risk, but they can also result in suppliers taking advantage of their preferred status to increase their own margins.

Storeroom Operations and Practices

The storeroom is where reliability of supply becomes tangible. A clean, organized, and well-run storeroom helps make maintenance faster, more accurate, and safer.

Best practices include:

- Control
  - Controlled access: prevent loss, theft, misplacement, and ensure issues and returns are accurately recorded.

- Efficiency
  - Kitting: prepackage all parts needed for a job to save time and reduce errors.
  - Transaction accuracy: every issue and return must be recorded properly.
- Service
  - 5S[2] organization: clean, labeled, and easy to find.

A good storeroom isn't just about control—it's about service. Maintenance should trust that when a part is listed as "available," it's really there, ready to go, and in good condition.

Data and Systems Integration

Many headaches come from poor data. Part numbers, bills of materials (BOMs),[3] and equipment hierarchies must be accurate and synchronized between your CMMS and ERP systems. If they aren't synchronized, planning accuracy and confidence both suffer.

Modern technologies like QR and barcoding, radio frequency identification tags (RFID), and even mobile scanning can dramatically improve accuracy and traceability. You can see what's been issued, to which asset, and how often—feeding valuable insights back into reliability analysis.

Controlling storeroom access is one way to ensure that technicians record what they take—*if they know that you know* they've been there, they are more inclined to record parts usage accurately.

Reliability of Spare Parts

Even from the same supplier, not all parts are equal—poor-quality or counterfeit components have caused major failures across industries.

Incoming inspection, supplier audits, and clear specifications can mitigate these risks. If you're using refurbished or repaired components, ensure the quality is verified and documented. The part's history matters—a "bad spare" can undo a lot of good maintenance. This is particularly important with repairable items that are being returned from the repair shop and reentering inventory. Having maintenance involved in these inspections, or even visits to the repair shops can go a long way to ensuring the quality of repairs. Treat every repaired or refurbished component as a reliability risk until proven otherwise.

Performance Measurement and Improvement

You can improve what you measure and there is much to measure with MRO supplies. Good MRO management includes clear KPIs, such as:

- **Stockout rate.** Parts unavailable when needed. You want this to be very low. Stockouts occur when parts are not there when needed.
- **Inventory turns.** How often stock cycles annually. You want this high, indicating that you have the right items that are actually being used. High inventory turns indicate healthy stock movement and accurate planning.
- **Service level.** Percent of parts requests fulfilled on time. Again, you want this high, but realize that having it perfect may drive inventory levels very high also. Like stockouts, this shows you have what's needed when needed. If inventory levels get high, consider reclassifying the parts. Classification based on your asset criticality, an indicator of risk tolerance, really matters.
- **Obsolescence rate.** Value of unused or outdated stock. Keeping this low requires good collaboration between planning and inventory management. If an equipment is no longer in use, identify the stock that was there for it, and either eliminate it or reduce its demand accordingly. Obsolete parts can easily clog your shelves while tying up working capital. Take care that many parts may be used in multiple assets. Lowering demand rather than removing from stock is best in those cases.

Regular reviews of these metrics can reveal slow-moving items, obsolete stock, or excessive rush orders—all signs of process improvement opportunities. The goal isn't to measure everything, it's to measure what drives better availability, fewer emergencies, and lower total cost.

AI-aided analytics can also be used with maintenance work orders, purchasing records, and inventory records, even if all three are in different "systems," to identify which parts are needed, which are not, which are overstocked, which should be added to stock, and even what may have been ordered but not used, likely hidden away in someone's stash. AI-driven insights are most powerful when they're shared across maintenance, stores, and procurement.

Collaboration and Governance

Ultimately, effective MRO management is a team sport. Maintenance, procurement, and operations must align around shared goals: asset uptime, reliability, and cost efficiency.

A practical governance model includes:

- Joint planning meetings between maintenance and materials.
- Shared KPIs (e.g., stockout rate, service level). Why not make inventory measures part of your maintenance manager's annual performance review and likewise make asset availability part of the review for your inventory and stores managers? When incentives are aligned, collaboration becomes the norm, not the exception.
- Clear ownership of decisions, with the appropriate authority for acting around stocking policies.

When everyone sees MRO as part of the reliability system, not just a storeroom, asset uptime follows naturally.

CHAPTER 9

COST, VALUE, AND RISK MANAGEMENT IN MAINTENANCE

Maintenance managers are constantly told to "control costs." Executives often see maintenance as an overhead—a necessary evil to minimize. Those who've lived through a major equipment failure, production loss, or safety incident know the truth: The cheapest maintenance strategy can be the most expensive in the long run.

This chapter is about seeing maintenance through the financial lens—not as a cost to contain, but as an investment in value, reliability, and business performance. Managing maintenance finances well is about finding equilibrium among cost, risk, and performance.

Maintenance: The Hidden Value Behind the Cost

Maintenance only shows up on the income statement as a cost. It's visible and measurable, and some of it is easy to target when margins tighten. Repairs cannot be avoided when equipment is "down," but proactive maintenance is discretionary; it's necessary but it can be deferred. While maintenance costs are visible, the *value* of maintenance doesn't show up in the same way—it hides in the things that don't happen. But the real story lies in what you don't see:

- The breakdown that didn't occur while the plant was running steadily

- The lost production that was avoided and the contractual commitments that were met
- The safety incident that was avoided and never made the news

These are "nonevents," but they are very real business outcomes.

Executives may celebrate a safety milestone, the result of consistent and careful attention to safety details. But they rarely even think to celebrate a quiet week of smooth production—yet that's the direct result of consistent, competent maintenance. Like maintenance, safety has a cost, and its results are intangible. The problem with maintenance is one of visibility: costs are tangible, value is not.

Imagine a diamond (see Figure 9-1): maintenance costs are the crown and table are visible above the wide part (girdle). That's what most of us see and pay attention to. In our businesses, we put a great deal of effort into measuring those costs. But the part below is still part of the same diamond and still has great value, but it's just not as visible. In maintenance, beneath that girdle you have a large mass of hidden value: available asset uptime, safety, reliability, environmental compliance, risk reduction, and customer satisfaction. Our accounting methods don't

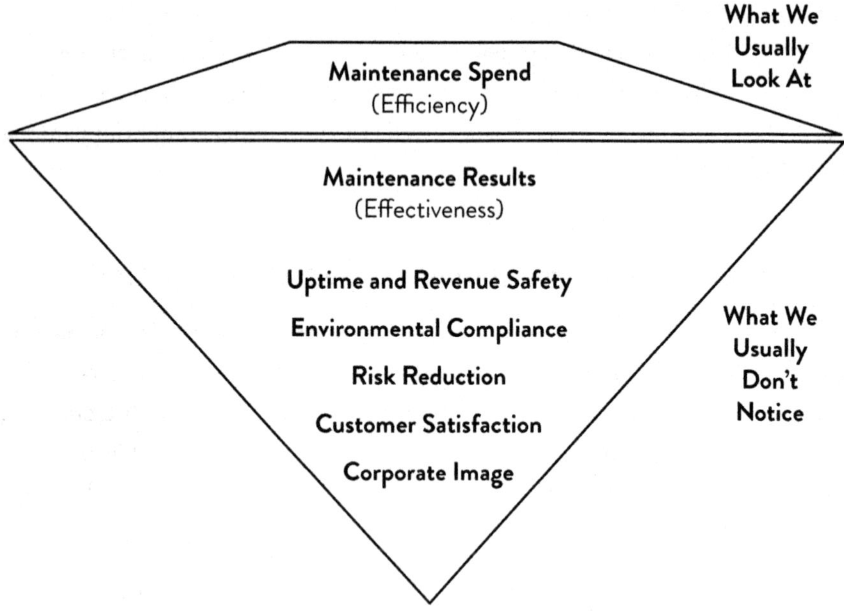

FIGURE 9-1 The Maintenance Value Diamond

measure those, but their value is still huge, and always a part of the whole. If we only look at the top, we are only seeing part of value that maintenance delivers.

Story: A pulp mill cut its maintenance budget by 15 percent to improve quarterly results, saving roughly $1.5 million of discretionary spending in the quarter. They eliminated all but regulatory-related proactive maintenance. That reduced spending on replacement parts, overtime, and contractors, and delayed some trades hiring. Six months later, a recovery boiler failed. Repairs were $10 million, and it caused a three-week shutdown, losing 54,000 tons of production, equivalent to $91 million in revenue. The unplanned loss of production cost roughly 50 times the "savings."

The lesson: Cutting maintenance costs is easy, a large part of it can be discretionary, but it can bite. The consequences appear later—usually far larger than the potential savings.

The lesson: Deferred maintenance is not a cost reduction—it's a transfer and usually an increase of risk to the future.

Maintenance Budgeting: Turning Plans into Financial Reality

A maintenance budget isn't just a spending plan—it's a reflection of how well you understand your work. If your maintenance plan is proactive and solid, your budget will make sense. If your plan is reactive or vague, your budget will be little more than a guess.

Different Budgeting Approaches

- **Top-down.** Management sets a target ("Cut 5 percent this year"), and maintenance tries to fit within it. It's simple, but it often ignores workload realities and asset condition. It's particularly inaccurate if the output or asset base being maintained is being increased or if past maintenance has been deferred.
- **Bottom-up.** The maintenance team builds a budget from known work—preventive tasks, repairs, inspections, overhauls. This is far more accurate, but takes effort and data discipline. It does not account for efforts that may go into making improvements.

- **Zero-based.** Every activity must be justified each cycle. It's great for uncovering waste but can be tedious for large operations. Requires a detailed plan for each asset or asset class. Very accurate if you've got the asset care plans (e.g., reliability-centered maintenance [RCM] outputs) on which to base it.
- **Activity-based.** The most insightful approach. Spending is linked to the *drivers* of maintenance cost—operating hours, hours of proactive work (PMs), materials required for PMs, asset criticality, failure rates, level of planning, and level of engineering support. This method helps connect financial outcomes to operational behavior, but may be difficult to match to accounting budgets, which are often focused on assets and broad spending categories, not activities.

Table 9-1 shows a typical maintenance cost breakdown:

TABLE 9-1 Categories of Maintenance Spending

Category	Description	Typical Budget Share
Labor	Internal maintenance staff	40% to 50%
Materials and spares	Parts, consumables	20% to 30%
Contractors and services	Outsourced work	10% to 20%
Overheads and support	Planning, tools, facilities	5% to 10%
Capitalized maintenance	Major rebuilds, upgrades	Variable

Tip: Track trends in each category—sudden shifts often signal process changes or hidden problems.

Best Practice

Tie the maintenance budget to the *production plan*. If production volume increases, asset utilization rises—so must maintenance needs. Too often, budgets are frozen while demand grows, leading to higher risk of failure and its attendant reactive repair work later. Note that among the activities you have repairs (must spend) and proactive maintenance (discretionary). Every dollar spent proactively typically avoids three in reactive repair.

Common Pitfalls

- **Across-the-board cuts.** They ignore differences in asset criticality and risk.
- **Flat budgets.** Using "last year + *x*%" ignores changing conditions like introduction of new assets or projects, and changes in planned outputs.
- **No distinction between fixed and variable costs.** Labor and support costs are largely fixed, but spares, overtime, and contractors vary with workload.

A strong maintenance budget is one that both finance and operations can understand—one that explains *why* the money is needed and *what value* it protects.

How Do You Know You Are in Trouble?

Quick Checklist: Signs of Unhealthy Maintenance Budgeting

- Deferred work backlog is growing.
- PMs are being skipped or shortened to save time. You might find phantom PMs, closed work orders with no labor or part charges.
- Breakdown costs are rising despite "savings."
- Spares stockouts are increasing.
- Engineering projects are being used to cover up chronic maintenance failures.
- Production losses are being accepted as "normal."

If these sound familiar, your maintenance budget may be under strain—and your risk is rising.

Life Cycle Costing: Seeing the Whole Picture

Upfront capital cost can be large, but often it is but a fraction of what you'll end up spending over the useful life of an asset. A well-maintained processing plant will last 60 or more years. It will cost you 3 to 5 percent of its original value per year just for maintenance, resulting in a spend over three times the capital cost.

That spending will vary year over year and there can be distinct peaks in spending at some points in an asset's life. For instance, large power transformers may have a large midlife refurbishment that effectively doubles their "life" at a cost of some 30 percent of a replacement.

The financial view of maintenance changes dramatically when you look at it over an asset's life—not just a fiscal year.

Life cycle costing (LCC) considers every dollar spent from purchase to disposal: design, procurement, installation, operation, maintenance, decommissioning, and site remediation. It reveals how decisions made early in an asset's life affect costs decades later. Note, too, that money spent up front on designing[1] the right maintenance program can have a huge payback. That design effort may raise the initial capital cost slightly, but the return is substantial.

Without it, the assumptions made by project engineers about future outputs may prove to be dramatically wrong. The "wrong maintenance" won't sustain the reliable performance that is often assumed.

Normally LCC takes account of the time value of money, meaning that a dollar today is worth more than a dollar next year. That is *not* done in Table 9-2 where there is a simplified example of an LCC comparison between two assets. One is more expensive up front (capital cost) but costs less overall at the end of the 10-year period. In that example, the more expensive asset to buy actually delivers greater value in terms of lower maintenance and energy costs every year.

TABLE 9-2 Life Cycle Cost Worksheet (Simplified Example)

Option	Purchase	Annual Maintenance	Annual Energy	Expected Life	Total LCC
A Low cost	$500,000	$60,000	$120,000	10 years	$2,300,000
B Premium	$600,000	$40,000	$90,000	10 years	$1,900,00

This is simplified and does not take in to account the discount rate for the value of money over time.

Result: Option B's higher upfront cost yields a 17 percent lower total life cycle cost.

Example

One large-scale nickel plant in Southeast Asia had been operating well below planned capacity because of flawed assumptions about reliable performance. It had been assumed that maintenance would just keep it going. It was running in a break-then-fix mode at 131 million pounds of output per year—plan was 150 million. A concerted effort implementing the concepts described in *Uptime*[2] with RCM increased production to 154 million in year one, 169 million in year two, and 175 million in the third.

The 1–10–100 Principle

As in product design, every dollar spent in design can save ten dollars in operations or a hundred dollars in failure costs. Conceptually that same principle applies in maintenance. Choosing a maintainable design, standardizing components, and ensuring access for service are small upfront costs that pay enormous dividends later. In fact, a program ensuring full operational and maintenance readiness executed during the design and build of a new asset might add 5 to 10 percent to the capital cost, but save ten times that in operating costs, and avoid a hundred times that in avoidance of risks to safety, environmental compliance, business continuity, and lost output.

Example

A municipal water utility compared two pump designs:

- The low-cost unit was $100,000 cheaper to buy but required frequent bearing changes and seal replacements.
- The premium model cost more initially but had half the maintenance burden and 20 percent better energy efficiency.

Over 10 years, the "cheap" pumps cost $400,000 more to own. Procurement saved money; operations paid the price.

I have visited hundreds of industrial operations in many industries, and I have heard many similar examples from operators and maintainers in all of them.

Procurement wins the quarter; maintenance sustains the decade.

Repair, Rebuild, or Replace

LCC also informs these common decisions. The rule of thumb: When the total cost of continued maintenance exceeds the replacement cost over the same time horizon—including the losses due to the ongoing risk of failure in the older asset—replacement is justified.

- Repairs are often less expensive than rebuilds, unless major damage renders the asset unrepairable.
- Rebuilds are more extensive and expensive, sometimes as much as replacement.
- Replacement can be complicated if you can't find a like-for-like option.

Simple guidelines can be helpful, but it doesn't take much effort to evaluate the options with uncomplicated spreadsheets. Don't ignore that replacement may also provide opportunity for additional capabilities and the need for modifications to accommodate it.

Best Practice

Include maintenance and reliability professionals in design and procurement decisions. The money they help save won't appear on a purchase order, but it will show up in sustained asset uptime and reduced operating costs.

When rebuild or replacement appears to be the best option, maintainers should involve operations and engineering in the decision. The opportunities for adding capability or the complications needed to accommodate a "not quite the same" replacement may warrant extra consideration of factors beyond purely maintenance considerations.

Cost, Risk, and Performance Equilibrium

Every maintenance decision sits inside a triangle having cost, risk, and performance as its three sides. To sustain a constant area (balance), the sides must remain proportional. Shift one side and the others are affected.

- **Cost.** What you spend on people, materials, and services

- **Risk.** The probability and consequence of failure (safety, environment, production, compliance)
- **Performance.** The reliability and availability your operation demands

Each impacts the others, and we want to maintain each within tolerable limits.

The key is finding the right equilibrium for your business context:

- If you chase the lowest cost, you'll raise risk and lose performance.
- If you pursue perfect reliability (lowest risk), then your costs will very likely skyrocket.
- If you want maximum performance, you'll be risking breakdown and costs will likely go way up.

The *sweet spot* is where total value—performance minus risk-adjusted cost[3]—is maximized. Table 9-3 shows a method used by many higher performing companies to achieve that. They look at the severity of consequences that are likely to be realized if the asset fails, and they consider how often that is likely to occur. In the lower right corner, for example, you can see that a rare event with catastrophic consequence (e.g., a fatality or a major loss of production) will still be treated as a highly critical asset.

TABLE 9-3 Risk Matrix Example

	Consequence →	**Low**	**Medium**	**High**	**Catastrophic**
Probability	Frequent	Medium	High	Very High	Extreme
	Occasional	Low	Medium	High	Very High
	Rare	Very Low	Low	Medium	High

Use: Rank asset risks and align maintenance priority accordingly. High-consequence, high-probability risks deserve proactive focus and investment.

Defining Tolerable Risk

Executives must define what level of risk the business can tolerate (i.e., live with). They must also recognize that "zero" risk is realistically unattainable. Risk can be an emotional topic that is difficult to understand fully. It is not taught particularly well in business or even engineering programs. Managers often view risk as "bad" and prefer to avoid it.

In practice, many managers prefer to accept the status quo because it is known, rather than make a decision that might actually improve the situation. The unknown is often perceived as high risk, even if the opposite is actually the case. Making no decision is a decision, and it can actually be riskier than accepting the new risk associated with efforts to improve on the status quo.

Once risk tolerance is defined or at least understood conceptually, the maintenance managers then align their strategies to meet that tolerance—no higher, no lower. This is where criticality analysis and risk matrices become essential tools. They are helpful in quantifying relative risks and allocating appropriate resources to deal with them.

For example, if a business wants a low risk of production loss, it will require a proactive maintenance program to lower that risk. Critical assets, the ones with the greatest potential negative impact on production, should get very careful attention in the proactive maintenance program, its timely and careful execution (doing the maintenance the right way), and in terms of what maintenance gets done. If not already done, a method like RCM would be the best option for defining that program. If performance today is within tolerable limits, then perhaps no action is needed on improvement, but if performance isn't satisfactory, decisions are needed (and likely funding) on what to do to improve it.

Example

A refinery deferred corrosion inspections to save $2 million. A year later, a heat exchanger leak released hazardous vapor, forcing a 10-day shutdown and a $20 million loss. The cost, risk, and performance triangle had clearly tipped too far toward cost savings, increased risk, and resulted in zero performance for the duration of that shutdown.

The consequences of ignoring the equilibrium are not theoretical.

Key Lesson

Balanced decision-making demands visibility into both probability and consequence. Tools like reliability modeling or even simple failure history tracking help quantify these trade-offs in business terms. Emotionally driven decisions about taking no risks and blindly lowering costs can produce very unsatisfactory business outcomes that are generally avoidable and predictable!

Communicating Maintenance Value for Executives

Maintenance managers often speak in technical terms—MTBF, downtime, PM[4] compliance. Executives speak in financial terms—revenue, margin, EBITDA,[5] ROI,[6] risk exposure. Translating between the two is a critical leadership skill. Bridging these vocabularies is a leadership act.

Reframing Metrics

- Availability → Production capacity
- Downtime hours → Lost revenue or output
- Maintenance cost → Cost of asset ownership
- Reliability improvement → Risk reduction or margin protection

When you express maintenance results in business language, the conversation changes.

Example

A reliability improvement program reduces downtime by 2 percent. In technical terms, that's a small gain. But in a $200 million production line, that 2 percent translates to $4 million in recovered throughput. Suddenly, maintenance isn't a cost—it's a profit enabler.

Dashboarding for Executives

While the maintenance manager needs to focus on processes, the executive is much more interested in outcomes. A good dashboard answers one question: *"What value did maintenance deliver this month?"* What executives want to see will include:

- Downtime cost avoided[7]

- Risk exposure reduced
- Reliability ROI (value of asset uptime ÷ maintenance spend)
- Cost of unreliability (lost opportunity from failures)

When executives see these numbers consistently, they can start asking better questions—and making better investment decisions. Table 9-4 is an example of what a value-focused table of metrics can look like. It shows costs and how you are progressing against budget, as well as results in terms of availability, downtime cost avoidance and safety performance and calculates a return on the reliability investment (maintenance spend). Below it is a brief explanation showing the relationship between spending saved and availability not being achieved. Is there any doubt as to where attention should be focused?

TABLE 9-4 Maintenance Value Dashboard

Metric	Month	YTD	Target	Business Value
Availability %	95.5	95.6	96.0	+$2.0m increased throughput per 1% Av. Currently below target.
Maintenance Budget Saved $m	1.2	9.8	10.0	Budget is $10m/month. Within plan.
Downtime Cost Avoided $m	0.7	5.2	>0	Production protected. Achieving target outputs.
Safety Incident (Count)	0	1	<=1	Compliant.
Reliability ROI % = Overall Benefit / Maintenance Spend	9% ($900/ $10m)	17.5% ($15.8m/ $90m)	>20%	Strong performance. Trending to 21% per year.

Maintenance budget is $120m/year. Figures are for month 9.
Spend is down by deferring PMs, but that likely caused Av to underperform.

Show Table 9-4 to an executive, and you'll have their attention.

Integrating It All: A Framework for Maintenance Investment Decisions

Managing cost, value, and risk effectively requires structure. Here's how to bring the principles of this chapter together in practice:

- **Define the objective.** What are we protecting or enabling (e.g., throughput, safety, compliance)? Quantify these.
- **Identify constraints.** Budget limits, manpower, asset availability, or risk tolerance. Beware that an arbitrarily low budget limits manpower and effort, potentially lowering availability while increasing risks.
- **Evaluate alternatives.** Preventive, predictive, redesign, replace, or run-to-failure. The best tool for making these decisions is reliability-centered maintenance (RCM).
- **Quantify costs and risks.** Estimate probability, consequence, and financial impact of each option.
- **Select the optimal solution.** The lowest life cycle cost that meets performance and risk objectives.
- **Monitor and adjust.** Track outcomes with cost, reliability, and risk metrics.

This process moves maintenance from "spending decisions" to "investment decisions." It involves finance as a partner, not merely a gatekeeper.

Key Takeaways and Executive Insights

- Maintenance is a value enabler, not a cost center.
- A well-structured budget reflects a well-defined maintenance program.
- Life cycle thinking prevents short-term savings from turning into long-term losses.
- Optimization—not minimization—of maintenance cost delivers sustainable performance.
- Translate reliability into visible business value.

Insight: When finance sees maintenance as an investment—and maintenance learns to speak the language of value—the business achieves reliability excellence that lasts.

Maintenance is financial stewardship in disguise. Every wrench turn, inspection, and replacement part is a small investment decision with business consequences.

Those who understand how cost, risk, and performance connect are not just maintaining assets—they're protecting enterprise value.

PART III

EXECUTIVE STRATEGY—MAINTENANCE AS A BUSINESS DRIVER

Part I laid the foundation for modern maintenance and reliability—its evolution and guiding principles. Part II covered the tactical execution. This section, Part III, looks at maintenance as a strategic business discipline.

If you've got physical assets, you have maintenance. Most organizations focus almost exclusively on the tactics (Part II) and keeping things running: The lights are on, but they are not moving forward. They are surviving but not thriving. Some have a deeper understanding of the foundations (Part I) and use that to enhance their tactics, improving performance. Indeed, many consulting services and software providers still focus primarily on tactical improvements, often solving one problem at a time. It's incremental and tends to be slow paced, often to avoid overloading already lean and somewhat overworked staff.

Improvements are likewise incremental and often fall far short of the real potential. Consider the last computerized management system installed and commissioned for maintenance: Did it deliver those promised savings and reliability improvements? Most likely not.

The real reason: It was treated as *the* solution—something to ease today's pain, not to transform tomorrow's performance. The improvement was focused on what was wrong, not on the root cause nor improving the business performance overall. There would have been a business case with an attractive return on investment that appeared to provide business benefit, from the software (or whatever "solution" you have in mind), but all that had to take place to deliver that full benefit wasn't done; it rarely is. One could imagine it was overlooked, but in truth, it was expensive. Too expensive to include in the business case for fear of damaging the ROI needed to justify the project. The tactically immediate need for the system ignored the full business reality. Sadly, such tactical considerations and actions are rampant in the siloed way we manage our businesses.

Information and Decision Support

Too often, we expect technology to deliver results simply because it is implemented. The truth is that systems—whether they are CMMS, EAM, or predictive analytics platforms—are only as powerful as the quality of the data and discipline behind them. Strategic decision-making depends on trustworthy information about asset condition, performance, and costs.

Executives need that insight to guide capital allocation, risk mitigation, and production planning. Good data turns reactive maintenance into predictive management. Without it, decisions are made on assumptions, not facts—and the organization drifts back into tactical firefighting. The real value of digital transformation in maintenance lies not in technology itself, but in the clarity and confidence it brings to decision making.

To break free of the tactical siloed thinking requires greater awareness at the executive level about the value that physical assets deliver, how they deliver it, and what we must do to sustain that value.

Strategic Asset Management

To truly elevate maintenance to a business-driving level, we must think in terms of *asset management*—the coordinated activity of realizing value from our physical assets. While maintenance is about sustaining function and reliability, asset management is about aligning those efforts with the organization's broader goals: profit, risk control, sustainability, and long-term viability.

Standards such as the International Standards Organization (ISO) 5500x describe this alignment as the "line of sight" between corporate strategy and day-to-day asset decisions. It's not a bureaucratic exercise; it's about ensuring that every dollar invested in our assets delivers measurable business value. When executives view maintenance through that lens, it becomes clear that reliability is not a technical goal—it's a strategic one.

As discussed in earlier chapters, value realization is not just about cost efficiency; it's about aligning asset performance with strategic intent.

CHAPTER 10

THE STRATEGIC IMPORTANCE OF PHYSICAL ASSETS

What We Want

From a business perspective, we truly want our physical assets to be:

- Producing at peak capacities to maximize revenues
- Maintained the right way to enable peak capacity operation
- Able to last long, in order to minimize the need for capital replacement in the near term
- Chosen to minimize total life cycle cost to our business, considering capital and operating costs, performance, and risks
- Safe to operate and perform within environmental regulatory constraints

Yet time and again, organizations struggle to sustain all these outcomes at once.

Maintenance intelligence is not just operational, it's financial.

Linking Maintenance to Capital Planning

Reliability data gathered through maintenance programs provides a powerful lens for capital planning. Knowing the true condition and performance of assets allows better timing of capital replacements, avoiding both premature renewals and catastrophic failures.

One customer, a wastewater treatment operation, is looking at all critical assets that are slated for replacement to evaluate condition, determine whether refurbishment is needed (and to what extent), or confirm end-of-life. So far in just two years, they've avoided $30 million in capital replacement, freeing up that money for other program needs.

A sound renewal strategy is not simply about deferring costs—it's about making informed investment decisions. When reliability and condition data is integrated with financial planning, we can see where capital should be spent to sustain performance and minimize life cycle cost. In that way, maintenance insight directly shapes the capital strategy of the business.

Shooting Ourselves in the Foot

The push for high capacity, "sweating the asset," often leads to excessive breakdowns, increasing risks to safety and risks of environmental noncompliance. If those breakdowns and discontinuities in production are substantial, insurance premiums to cover business continuity will be higher. Safety records will be worse, and workers' compensation insurance rates will be higher, and you may be paying fines and potentially facing criminal charges. If you've suffered environmental noncompliance, you will likely pay fines and put your license to operate at risk.

If we don't look after our physical assets correctly, we are opening the door to a range of undesirable consequences that can seriously impact our business. The strategic importance of physical assets lies as much in what they *prevent* as in what they *produce*. We can see production and measure revenue, but we can't see risk and its consequences until it is too late.

Most businesses depend heavily on the performance of their physical assets to perform many routine and production-related functions. Physical assets in the form of computers and networks help us manage the business and its production and service delivery processes. By automating, we speed up the processes and make them consistent, increasing both quantity and quality of what we do. We also remove a lot of human intervention, saving on labor costs.

We invest in those physical assets because they can do things faster, cheaper, more consistently and more accurately than we can do them by hand or with our minds alone. Businesses are competitive by nature and to be the most profitable we need to produce or deliver more, at a lower cost than our competitors. Government agencies are not driven by profit, but they are answerable to taxpayers and meet demands for services to be delivered at the lowest practical cost.

The consistent theme, regardless of sector, is "more for less," where "more" is all about what you deliver, and "less" is about all the costs associated with producing it.

Cost is about money, in any form. The "more" is about output—goods or services.

Although we understand life cycle costing and the benefits of proactive maintenance, we seldom spend more capital to ensure minimum operating and maintenance costs during the operational phase of the life cycle of the asset. Project economics tend to discourage spending anything extra. That's how our operations are full of nonstandardized equipment, our fleets are full of vehicles from a range of manufacturers, our parts inventories are bloated with a whole lot of functional duplication, and our training needs to consider variations in equipment all designed to do the same thing.

These inefficiencies are not just technical inconveniences—they directly affect profitability. Every shortcut in asset care leaves a hidden liability on the balance sheet.

Maintenance Contributes to Profit

Profitability means having the greatest margin, and we obtain that by having the highest possible output (assuming we can sell it) at the lowest possible cost. We also need to obtain that safely and without violating any environmental or other regulations.

Margin is represented in Figure 10-1. It's the result of revenue minus total costs. If production is below the breakeven point, you are losing money, so production level must be kept to the right. The farther to the right, the better: that increases the margin and hence profits.

Doing the right maintenance the right way has a positive effect on operating income. Doing maintenance the right way lowers costs. Doing the right maintenance increases availability, thereby increasing output and revenue. The right maintenance also reduces the amount and cost of maintenance needed, so the breakeven point moves further left and margin potential grows at all points to the right.

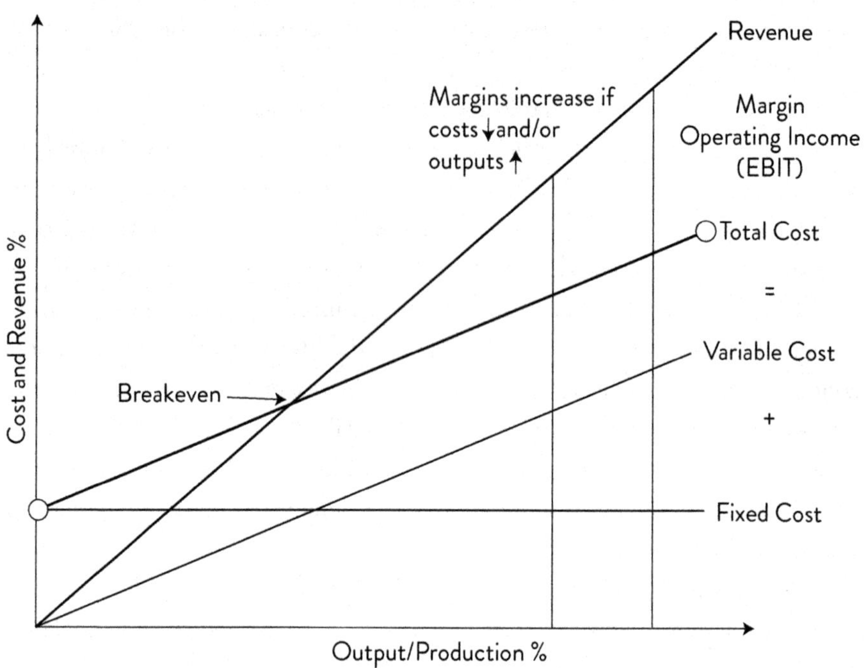

FIGURE 10-1 Impact of Maintenance on Margin and Breakeven Point

Safety and Environmental Compliance

Environmental compliance entails containment and controlled emissions, all within stated limits. Environmental infractions occur as a result of equipment and system breakdowns. They fail to contain, or fail to control—two functions of a lot of our equipment. If we minimize the probability of those failures, we also lower the risk associated with failure to comply with environmental regulations. We need the right maintenance to produce the reliable performance of our assets to sustain those containment and control functions.

It has been shown, time and again, that safety incidents are lower in reliable operations.[1] Safe sites are reliable and cost-efficient. As Ron Moore put it, in a course he was teaching to one customer, "if you want to get serious about safety, then get serious about reliability."

All too often, companies take an approach to safety that emphasizes protective equipment and following procedures, but they fail to deal with the behaviors that give rise to accidents. Safety isn't just about wearing appropriate personal protective equipment (PPE); it should also be about lowering the risks that give rise to unsafe situations that will require the PPE to do what it is designed to do.

Like wearing protective gear when riding a motorcycle—it minimizes harm, but true safety comes from disciplined behavior and a well-maintained machine.

Maintenance and Sustainability

Maintenance and reliability also have a direct link to the sustainability and environmental, social, and governance (ESG) agenda. Reliable assets consume less energy, produce fewer emissions, and generate less waste. Keeping equipment operating at optimal efficiency is one of the most practical forms of carbon reduction available to industry today.

One U.S. Department of Energy study and another by Portland Energy Conservation Inc. show that up of 20 percent energy savings can be achieved by eliminating parasitic losses caused by poor or lacking maintenance.

In most production environments, a 20 percent reduction in electricity and fuel costs justifies targeted investment.

From an ESG standpoint, a well-managed asset base demonstrates stewardship, showing that the organization looks after what it owns, minimizes waste, and extends useful life. These are not just compliance goals; they are reflections of corporate responsibility and increasingly, competitive differentiators in the marketplace.

Reliability as the Foundation of Resilience

In recent years, global events—from supply chain disruptions to extreme weather—have revealed the fragility of many operations. Reliability and maintenance play a central role in building resilience. Reliable assets keep the organization productive and compliant under stress. Well-managed spares and proactive planning protect against supply chain shocks.

Maintenance excellence is therefore not just a matter of efficiency—it's a cornerstone of operational resilience and risk management. The organizations that weather crises best are those that have made reliability a strategic priority long before it was tested.

Reliability doesn't just support strategy—it protects it when conditions change.

The Whole Is Greater Than the Sum of the Parts

There are benefits in doing better at planning, at scheduling, at measuring your KPIs, at implementing a computerized management system, at implementing lean concepts, and so on. If you do them all together, you get more of the benefits of increased efficiency. If you are doing the right work, you also benefit from the effectiveness of a well-designed maintenance program. Each part contributes, and the effects are far greater than any one of those initiatives on its own.

Planning provides the know-how to do both the proactive and repair work, and what we need to do it efficiently. We can execute the work more quickly, with greater precision and no delays.

We may find something that wasn't in the plan, but we are well ahead of having no plan at all.

Work execution is less expensive—about one-third the cost of unplanned work (on average). But if you are doing too much of the wrong work, you don't get that full value.

By ensuring we have the resources that our plans specify before we schedule the work, we add confidence in achieving the schedule in your maintainers and operators alike. Scheduling will ensure we get the more important and urgent jobs done first, with minimal delay, while maximizing the utilization of our workforce.

The ability to stick to that schedule is impacted by the quality of plans and the frequency of schedule-interrupting breakdowns.

Having the right proactive program results in fewer breakdowns—longer productive runs. That program will be largely predictive—executed without disruption to production. When we find problems, we have the lead time to stage parts and resources and perform the proactive repair at a time that minimizes impacts of an otherwise inevitable failure. To achieve that, we need disciplined follow-up on those predictive findings and job plans at the ready.

The right maintenance program is dominated by proactive work. Since most failures come with a warning, we have time to prepare without last-minute rush.

The proactive work is all done at known frequencies—demand for materials is known with precision. We also know that a proactive repair will be needed eventually, so parts forecasting is also enabled. We leverage our knowledge of our equipment reliability to forecast demand; probabilistic algorithms determine correct stock levels to meet it.

Proactive work identification and planning communicate with inventory management to make that work. Stores and procurement must be able to respond to demand in a timely manner. Even if part lead times are small, cumbersome procurement processes create inefficiencies that slow the velocity and increase the risk of plant outages.

The right proactive program can have a very positive impact on stores and supply of those parts. We effectively turn much of the inherently random demand for parts and materials into something far more predictable and manageable, as depicted in Figure 10-2.

While we can reduce working capital tied up in stores, and lower procurement costs, our maintainers must be disciplined. They must refrain from bypassing those processes with direct purchases, rush ordering, and buying an "extra" to stash away. They mean well but mess up that integration. Shadow inventories and direct purchasing distort consumption data, compromising the very foundation of good inventory management.

When these disciplines reinforce one another, maintenance becomes an orchestrated system, not a series of isolated tasks.

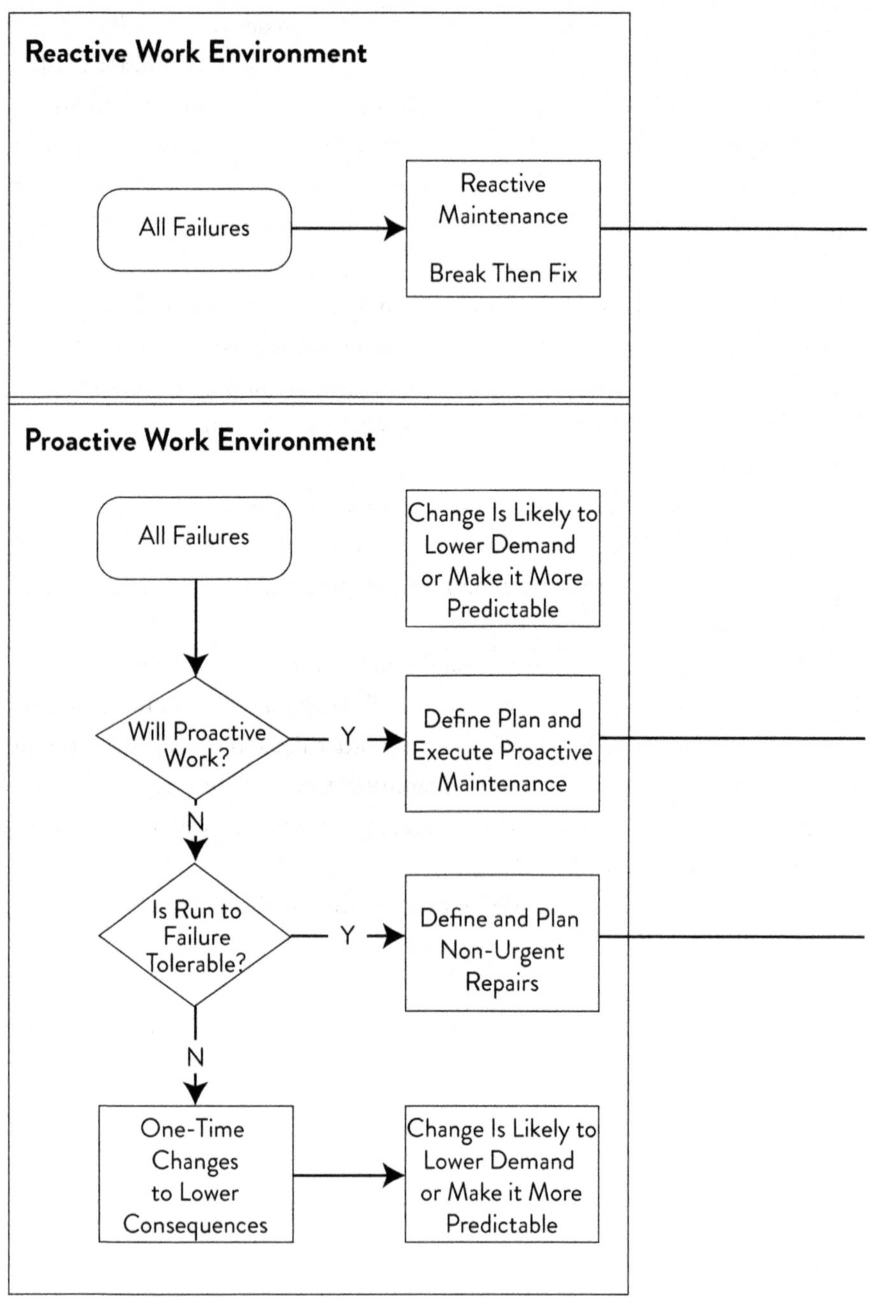

FIGURE 10-2 Integrated Work Identification, Planning and Materials Processes

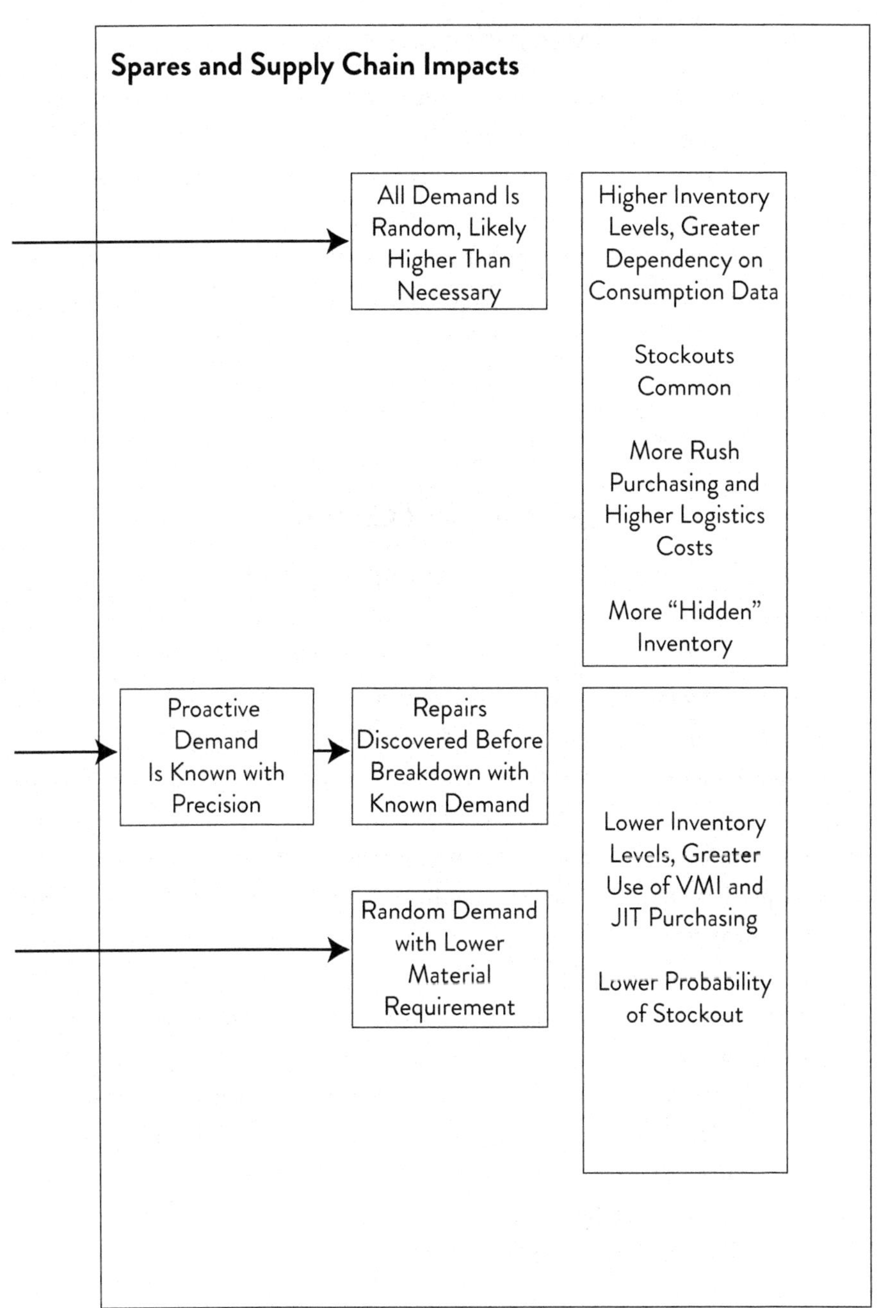

FIGURE 10-2 Integrated Work Identification, Planning and Materials Processes *(continued)*

Competitive Advantage

There are benefits that accrue with each improvement effort, and they multiply if combined. Consider the list of those possible improvements like the list of ingredients that comes with a recipe. It's not a list to cherry-pick from; you need every ingredient, and you need to know how to combine them.

Simply throwing them all together with no plan or guidance can create confusion and even set you back. Part IV of this book deals with how to pull the ingredients together correctly. It completes the recipe.

Of course, even the best "recipe" won't succeed without the right leadership mindset to guide and sustain it.

Leadership and Culture

Achieving strategic leverage from maintenance begins with leadership mindset. When executives treat maintenance as a cost to be minimized, that's exactly what it becomes. When they view it as a value generator—a function that enables throughput, safety, and reliability—it becomes a competitive advantage.

The culture must shift from "fixing things that break" to "designing out failure and managing risk."That shift won't happen through slogans; it happens when leaders measure, reward, and discuss performance in terms of reliability, risk, and value delivered. Cross-functional alignment starts at the top, with clear expectations that production, maintenance, supply, and engineering all share ownership of asset performance.

There is a need for collaboration among the various business functions. Maintenance, engineering/reliability, operations/production, stores/inventory management, and purchasing all need to be a part of this to maximize the benefit. When you get into using KPIs as performance drivers, you'll need human resources and finance involved to align compensation and reward schemes for the behaviors you need to encourage.

Leaders set the tone by asking not "What did maintenance cost us?" but "What value did reliability create?"

Measuring What Matters

This is where decision support (Chapter 7) meets accountability: the metrics must close the loop between technical performance and business outcomes.

Key performance indicators (KPIs) are useful only when they link directly to business outcomes. Metrics like mean time between failures (MTBF), maintenance cost, or schedule compliance have meaning only when they connect to productivity, margin, and risk reduction. Executives should focus on *value realization—tracking* not only the costs avoided, but also the financial gains from improved asset availability, energy efficiency, and risk mitigation.

Too often, organizations celebrate improved internal metrics without confirming whether the business itself has benefited. The goal of the maintenance manager is to "do maintenance better," but the executive's goal is to enhance the performance and profitability of the enterprise.

Truly effective measurement connects the workshop floor to the boardroom table.

Benefits

The benefit at a single site can be large—averaging 20 percent cost savings in maintenance, and easily ten times that in revenue gains. Doing this across multiple sites in a large organization increases the benefit overall and to the owners or shareholders.

In one mining company with six operating sites in three continents, we found over $90 million in cost savings and nearly $1 billion in potential revenue gains annually.

Other benefits include:

- Fewer safety incidents
- Greater compliance with environmental regulations
- Less risk of noncompliance with regulatory inspections
- Less variability due to breakdowns, increases your ability to meet contractual delivery commitments
- Increased output and margin
- Reduced requirement for working capital tied up in stores

- Reduced need for capital replacements because existing equipment lasts longer
- Less risk in safety, environment, and business continuity can lower business insurance premiums
- A less chaotic working environment is nicer to work in—employee retention is less of a problem
- The reduced workload from being proactive lowers demand for overtime, contractor labor and even hiring of skilled trades (who are very hard to find nowadays)
- Increased "wrench time" from good work management increases trades' job satisfaction and reduces attrition

Each of these benefits compounds across sites, turning maintenance from a cost center into a strategic value-multiplier.

Getting there requires that you begin with an understanding of the value generated by maintenance and reliability from that strategic perspective. In a single site, it begins with the general manager (GM). Across a portfolio of sites that understanding is also needed both in the GMs and their chief operating officer (COO).

You may already have a corporate center of excellence, or some equivalent, responsible for your physical assets, but these groups are often toothless. They are often staffed by strong site practitioners, but lack the authority to drive enterprise-level change; they become troubleshooters rather than owners of the reliability agenda.

As an executive, you are now seeing that a lot of value is still there to be achieved. Ask, "Why haven't we achieved it?" The answer is seldom a lack of capability; it is usually a lack of sustained executive sponsorship.

Part IV will speak to how this all comes together.

Recognizing and managing physical assets strategically isn't optional—it's the foundation of long-term success.

Summary: The Strategic Importance of Physical Assets

Physical assets are more than the tools of production—they are the foundation of business performance, safety, and resilience. When managed strategically, they link day-to-day maintenance to capital planning, risk control, and profitability.

Maintenance done "the right way" not only prevents failures but improves margins, extends asset life, and reduces environmental and safety risks. Reliability data guides smarter investment and renewal decisions, aligning asset performance with corporate objectives and life-cycle value.

Reliable operations are safer, more sustainable, and more resilient. They consume less energy, generate fewer emissions, and sustain productivity through disruption. These outcomes depend on leadership mindset, disciplined collaboration across functions, and measurement that ties maintenance performance to business results.

In short, the strategic importance of physical assets lies in their ability to create and preserve value. When reliability becomes a leadership priority, maintenance evolves from a cost center into a competitive advantage—and the organization moves from surviving to thriving.

CHAPTER 11

MAINTENANCE AND ENTERPRISE RISK MANAGEMENT

Maintenance is often seen as a tactical function: fixing things when they break, scheduling work orders, and keeping the lights on. In the boardroom, maintenance decisions influence operational continuity, brand integrity, and even shareholder value. From preventing unplanned downtime to ensuring compliance and safety, maintenance is on the front line of enterprise risk management (ERM).

Increasingly, ERM is among the top concerns in board thinking. To address it, there is an international standard, ISO 31000:2018. It defines risk management through three components: principles (value creation and protection), a framework (leadership and integration), and a process (for assessing and treating risks).

Understanding Risk in Maintenance

Every operation carries risk. Some risks are obvious: a critical pump fails, production stops, and revenue is lost. Others are more subtle: regulatory noncompliance, environmental incidents, or latent safety hazards. Maintenance teams are uniquely positioned to identify and mitigate many of these risks before they materialize.

Think of risk in three broad categories relevant to maintenance:

1. **Operational risk.** Unplanned downtime, production losses, degraded performance

2. **Compliance risk.** Regulatory, contractual, or standards violations
3. **Safety and environmental risk.** Injuries, accidents, or pollution from asset failures

The consequences of ignoring any of these can be severe, not only financially but also reputationally.

Maintenance isn't just about fixing machines; it's about protecting the enterprise from these cascading impacts.

Maintenance as a Risk Management Function

Integrating maintenance into the ERM framework transforms it from a cost center to a strategic risk management function. Within the ISO 31000 framework, shown in Figure 11-1, maintenance contributes at every stage of the risk process:

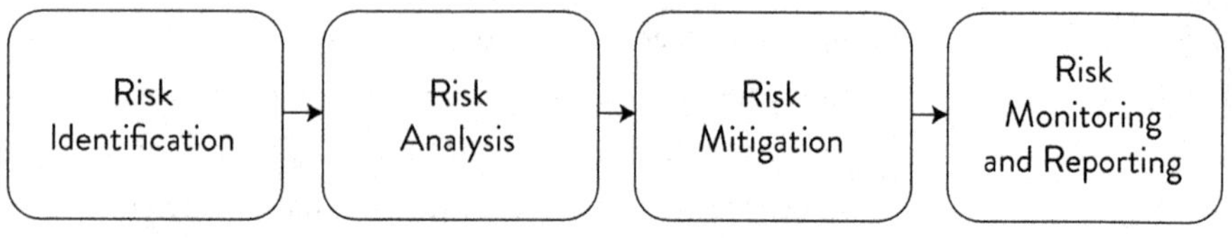

FIGURE 11-1 Risk Management Process

- **Risk identification.** Maintenance teams have firsthand knowledge of equipment vulnerabilities, failure modes, and operational weaknesses. Their insights feed directly into risk registers and assessments.
- **Risk analysis.** Understanding the likelihood and consequence of failure allows prioritization of maintenance resources. Not every failure carries equal consequence; the most critical assets warrant proportionate attention.
- **Risk mitigation.** Preventive and predictive maintenance programs directly reduce the probability of failures, while contingency planning reduces their impact when they occur.
- **Monitoring and reporting.** Maintenance KPIs—like availability, mean time between failures (MTBF), and compliance metrics—become part of the enterprise's risk dashboards, giving executives real-time visibility into operational risk.

Asset Criticality

Once risks are identified, the next question is where to focus. Asset criticality analysis (ACA) answers that.

ACA is a foundational process (see Figure 11-2) within an ERM framework that helps organizations identify, evaluate, and prioritize their key assets to better allocate resources for risk mitigation. It begins with a comprehensive inventory of assets, which can include physical items like equipment and facilities, as well as intangible ones such as intellectual property, data, and personnel. Once you have determined the most critical assets, you can determine how best to identify the most appropriate risk mitigation.

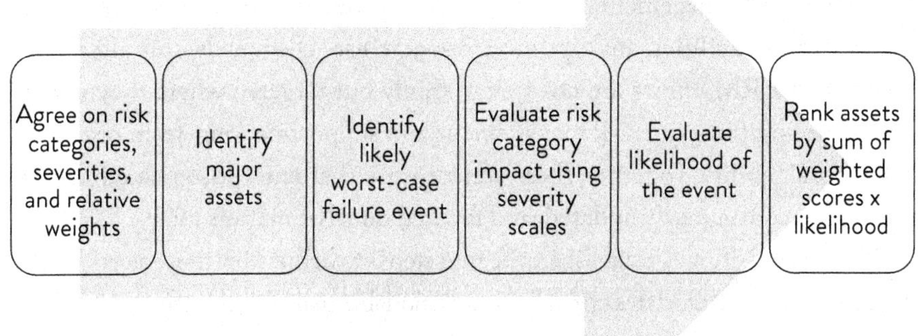

FIGURE 11-2 ACA Method

Criticality is assessed based on multiple criteria, such as the asset's impact on business operations, financial performance, regulatory compliance, customer satisfaction, and overall strategic objectives. For instance, a scoring system might be employed—often qualitative or quantitative—where assets are rated on scales for consequences of failure, relative rankings of those consequences, likelihood of threats, and recovery time objectives. This analysis integrates into ERM by enabling selection of high criticality assets for deeper analysis efforts such as reliability-centered maintenance (RCM). Strategically, this helps avoid significant and cascading effects on the business. It enhances resilience by informing strategies like maintenance programs, insurance coverage, contingency planning, and cybersecurity investments.

In practice, ACA operates iteratively within ERM. It involves cross-functional teams from IT, operations, maintenance, engineering, finance, and compliance to ensure a holistic view.

ACA is usually applied to your maintenance significant assets. Those are the assets you'll easily identify as major parts of your production or delivery process (e.g., production systems, major static and rotating equipment, conveyances, mobile fleet assets).

Tools are available to aid the process; a common one being tailored spreadsheets. They enable dynamic updates as business environments evolve. By ranking assets—say, "high criticality" to "low," or from "mission-critical" to "nonessential"—your risk management team can prioritize controls and monitoring, reduce potential vulnerabilities, and optimize resource use. Ultimately, this alignment ensures that ERM efforts are not spread thinly but targeted where they matter most, fostering proactive risk management and supporting long-term organizational sustainability. Together, these steps ensure that enterprise risk efforts are data driven, continuously updated, and focused on what matters most.

With asset criticality established, the next step is to define how those assets should be cared for; RCM translates criticality into actionable maintenance strategies.

RCM and Risks

RCM is a systematic, structured process used to determine the most effective failure management policies for preserving asset functionality, while minimizing risks to business operations.

Often overlooked, RCM is an excellent fit within the corporate ERM framework. Combined with ACA, RCM targets your most critical assets: those with the greatest potential risks to the business. Figure 11-3 shows how RCM enables the Risk Management Process for risks that can arise due to failures of your physical assets. Furthermore, RCM considers all likely causes of failures, including human error—a significant contributor to many failures, industrial accidents and even disasters. RCM systematically addresses each potential failure—its causes, effects, and consequences—then identifies the most effective actions to reduce risk to tolerable levels.

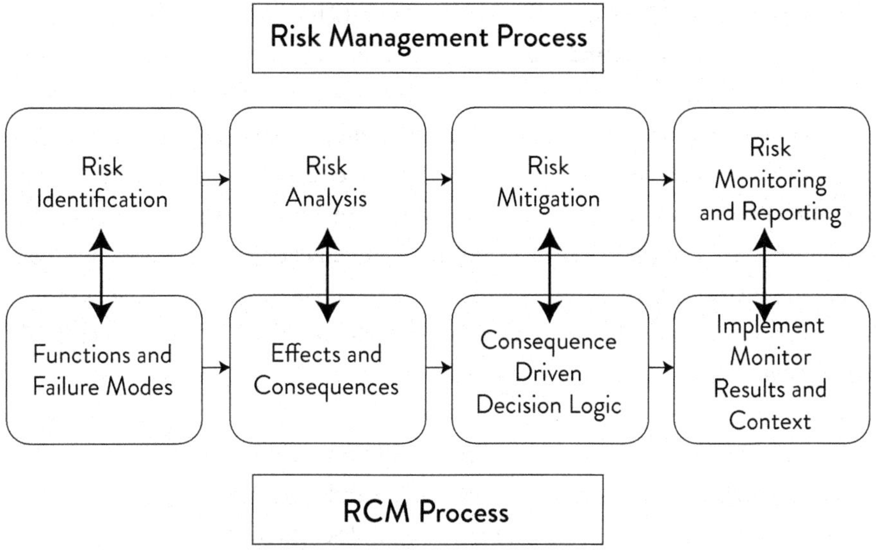

FIGURE 11-3 RCM Alignment with Risk Management

Example

A large wastewater treatment facility had experienced a significant release of biogas in their cogeneration plant. The release triggered alarms, drew complaints from nearby commercial and residential complexes, and required substantial effort to evacuate the explosive gas from the affected buildings. They were able to identify the location of the problem and added extra gas-monitoring equipment to provide earlier warning if it ever happened again, but they did not identify the cause of the release. Later, RCM analysis revealed the cause: an error by operators who didn't fully understand how the system worked. The addition of a new procedure and relevant training dealt with that human error, eliminating it for good.

The importance of RCM in ERM lies in its ability to optimize resource allocation, reduce unplanned downtime, and enhance operational reliability, thereby mitigating financial, reputational, environmental, and safety risks. By focusing on failure consequences rather than arbitrary schedules, RCM prevents overmaintenance of low-risk assets and undermaintenance of critical ones, leading to cost savings of 20 to 50 percent in many cases, while improving asset availability. Where used, it is often performed largely as a maintenance and reliability exercise, but it has far-reaching benefit.

Integrated into ERM, it provides actionable data for risk registers, informs capital investment decisions, and supports compliance with standards like ISO 31000 (risk management) and ISO 55000 (asset management). RCM is very sensitive to "operating context" and as such its results can be easily adapted to changes in business conditions. Business demand, changes in customer usage, changes in the operating environment itself, supply chain disruptions, or regulatory shifts can all be accommodated. Coupled with an active outlook for conditions that might impact on assets and their associated risks, and fostering a proactive culture, RCM helps transform maintenance from a cost center into a strategic enabler of enterprise-wide risk management and long-term sustainability.

Lesson learned: One major electric distribution utility experienced a major transformer explosion, causing the evacuation and need for temporary housing for all residents in a high-rise apartment complex. The utility had done RCM analysis and was following its recommendations. However, the operating context had changed, and no one thought to examine its potential impact on asset failure modes and how those assets were being maintained. In addition to densification in the urban environment (increasing load), the utility had recently introduced "time of use" billing. That changed the already cyclical load pattern into one with higher peaks and shallower valleys, impacting heating and cooling of the assets, eventually resulting in failure of some. Once recognized, the utility reviewed all its asset RCM analyses and made several substantial changes to its failure management policies.

Embedding Maintenance in Enterprise Risk Frameworks

To be effective, maintenance must not operate in isolation. It needs to be part of a larger ERM ecosystem:

- **Cross-functional collaboration.** Enterprise risks are rarely siloed. Maintenance must work with operations, engineering, safety, finance, IT, and compliance teams to understand and mitigate enterprise-level risk.
- **Data integration.** Asset data, inspection reports, and predictive analytics should feed into the organization's risk management tools. This ensures that risks are quantified, tracked, and prioritized. Incomplete maintenance data no longer suffices when maintenance assumes a strategic role.
- **Decision support.** Maintenance strategies should support risk-informed decision-making. For example, deciding whether to run equipment to failure, perform condition-based maintenance, or replace an asset early should all be grounded in risk analysis. RCM is the best tool available for that decision making.

Risk and Maintenance Integration

Reliability (the ability to continue performing desired functions) is at the very heart of managing assets and risks. Both are foundational to a strategic perspective on maintenance management.

Physical assets embody multiple potential risks—operational, compliance, safety, and environmental—each arising from specific failure modes. All of those are undesirable consequences of failures that can occur in any of our physical assets.

Figure 11-4 shows that hierarchy beginning with standards (at the bottom) all the way to activities in your operations.

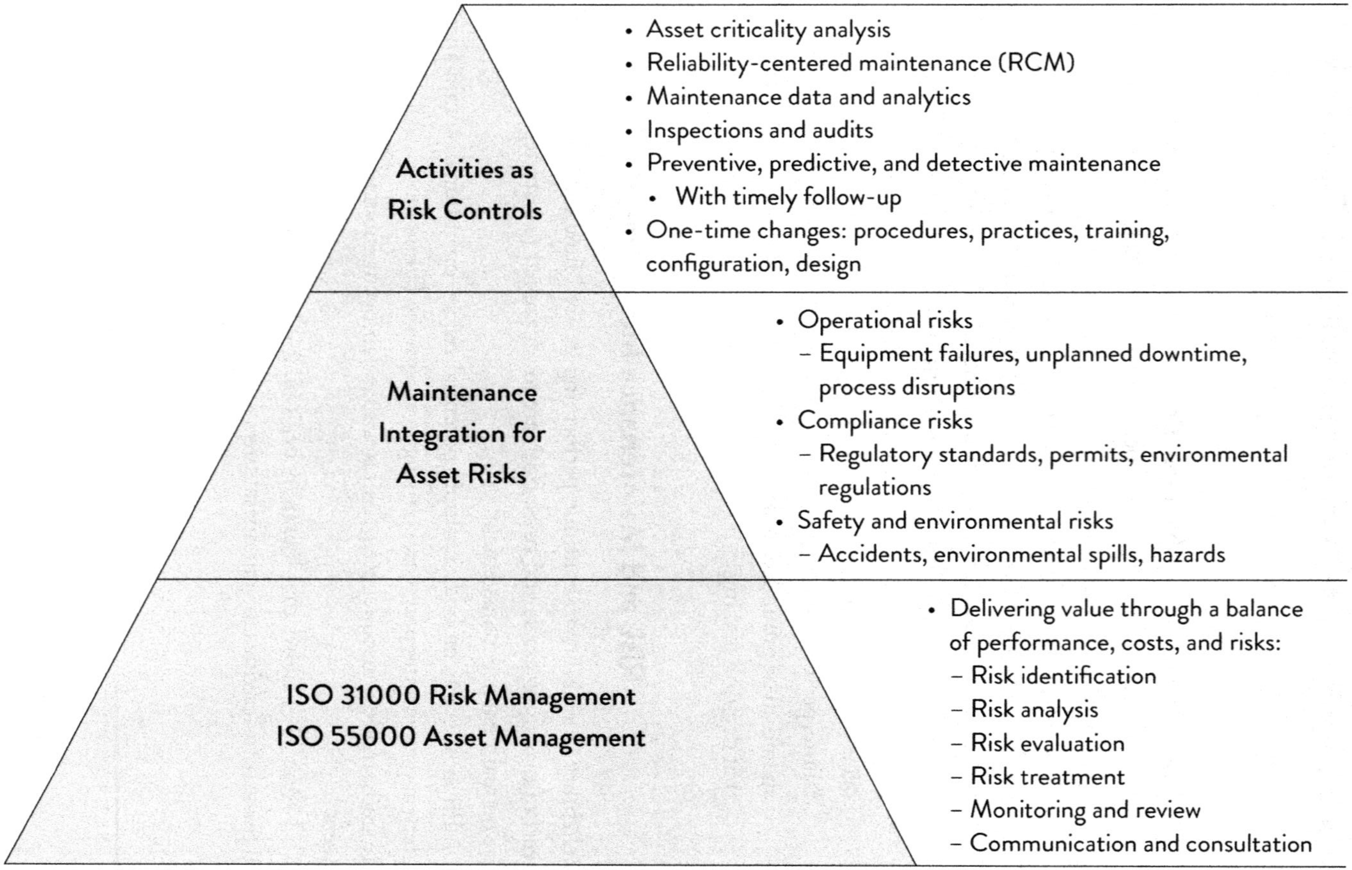

Activities as Risk Controls

- Asset criticality analysis
- Reliability-centered maintenance (RCM)
- Maintenance data and analytics
- Inspections and audits
- Preventive, predictive, and detective maintenance
  - With timely follow-up
- One-time changes: procedures, practices, training, configuration, design

Maintenance Integration for Asset Risks

- Operational risks
  - Equipment failures, unplanned downtime, process disruptions
- Compliance risks
  - Regulatory standards, permits, environmental regulations
- Safety and environmental risks
  - Accidents, environmental spills, hazards

ISO 31000 Risk Management
ISO 55000 Asset Management

- Delivering value through a balance of performance, costs, and risks:
  - Risk identification
  - Risk analysis
  - Risk evaluation
  - Risk treatment
  - Monitoring and review
  - Communication and consultation

FIGURE 11-4 Standards Lead to Activities

The ROI of Risk-Aware Maintenance

The payoff from integrating maintenance and risk management extends beyond compliance—it strengthens resilience and performance:

- Fewer unplanned outages and production losses
- Reduced safety and environmental incidents and compliance breaches
- Smarter investment in critical assets, aligning maintenance spend with enterprise priorities
- Improved confidence for executives, regulators, and stakeholders that risk is being actively managed

Key Takeaways

Maintenance is not just a cost or a support function; it is a critical enabler of enterprise risk management. Understanding and mitigating operational, compliance, and safety risks must be central to maintenance strategy. Integration with ERM frameworks, supported by data, cross-functional collaboration, and risk-informed decision-making, elevates maintenance from tactical to strategic.

CHAPTER 12

ASSET MANAGEMENT FRAMEWORKS AND STANDARDS

Why Frameworks Matter

When you talk to executives about "asset management," you quickly realize people mean different things. I've had countless conversations about this. What I hear from many is, "It's maintenance with a fancier name." From others, "It's a way to manage expenses and future capital requirements," while others (often in utilities or government) will see it as a corporate governance structure. In truth—it's all three—and more.

Asset management frameworks give us a common language. They bridge the gap between day-to-day maintenance work and the boardroom's focus on performance, risk, and value. Standards like the ISO 5500x series[1] don't tell you *how* to maintain an asset; they tell you *what good looks like* when your organization is managing assets as part of its strategy, for example, linking capital planning to reliability data. This is where those organizations finally "get it" with respect to maintenance and reliability.

It's about seeing your assets not as cost centers, but as value enablers: sources of revenue, customer satisfaction, and brand trust. That's a radical—and necessary—shift.

From Maintenance to Asset Management

Maintenance is about *doing the right things right* for an asset. Asset management asks a deeper question: "Are we doing the right things across the business?" It takes us out of tactics and helps us align maintenance and operations with business purpose.

In my maintenance management training classes, I show a diagram of nested dolls. Asset management is the outer doll, maintenance is the next one in, and reliability is inside that—at the heart. This visual helps executives grasp scope instantly. The diagram of boxes in Figure 12-1 depicts that same concept. Note that within the broad context of asset management we also have important interactions in human resources, training, supply chain, logistics, and so on. The three smallest boxes represent the three main parts of the *Uptime*[2] "Pyramid of Excellence."

In many ways, ISO 55000 formalizes what others and I have been teaching for years: seeing maintenance as one nested part of a much larger system.

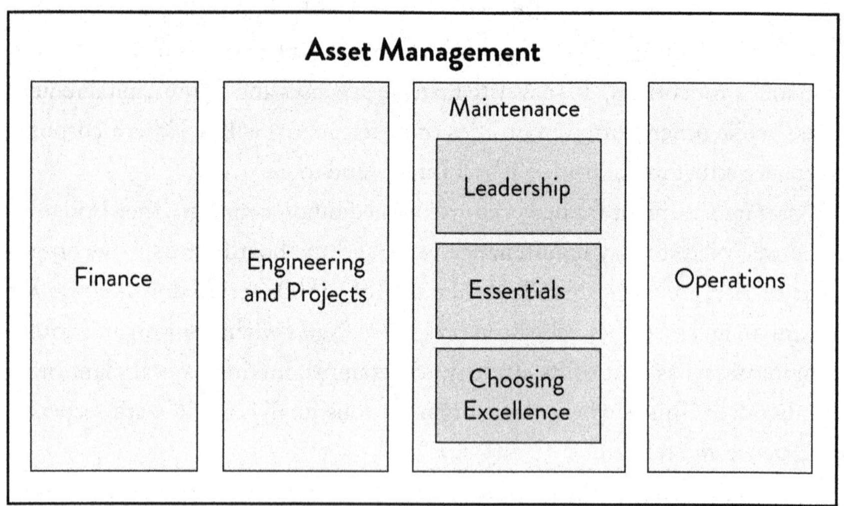

FIGURE 12-1 Maintenance and Its Components Are Part of Asset Management

Where maintenance plans optimize availability, asset management plans optimize value over the asset's entire life cycle—from design to disposal.

Table 12-1 shows how the conversation changes as you move from maintenance thinking to asset management thinking.

TABLE 12-1 Shifting Focus

Maintenance Focus	Asset Management Focus
Reliability of individual assets	Performance of asset portfolio
Maintenance costs	Total life-cycle costs (return on assets)
Work execution	Strategic alignment with corporate goals
Technical success	Business value and risk balance

Both are essential—but the higher you go in the organization, the more the conversation must pivot from reliability metrics to value creation and risk control. That is where frameworks become indispensable and that's exactly where frameworks earn their keep.

The ISO 55000 Family Explained

You've likely heard people reference "ISO 55000" as if it's one document. In reality, it's a family of three standards, supported by additional guidance materials. The first three are:

- ISO 55000—*Overview, Principles, and Terminology*
  Introduces key definitions and the "why" behind asset management.
- ISO 55001—*Requirements*
  The certifiable standard—what an organization must have in place. This includes all the "shall" statements against which an organization would be audited for certification.
- ISO 55002—*Guidelines for Application*
  Provides practical guidance for implementing the requirements.

ISO 55000 isn't prescriptive. It doesn't say, "You must plan maintenance this way." Instead, it asks:

- "Do your asset decisions support organizational objectives?"
- "Do you manage risks consistently?"
- "Is leadership actively involved?"

These questions can feel uncomfortable—and that's the point!

It's a management system, not a technical manual. And that's precisely its power—it connects reliability engineering with governance. That connection is often missing in organizations stuck at the "maintenance" level.

I've seen this play out in dozens of companies: Engineers take the lead, and executives quietly step back. In one utility I worked with, the CFO saw it as a cost-reduction tool until reliability engineers began leaving, letting HR know that they felt they were wasting their time making recommendations that were destined to be ignored.

Asset management has begun to show up in legislation and regulatory requirements, but not always with any reference to the standards. In those cases,[3] we've observed various interpretations, mostly under financial leadership, that miss some or even much of the intent as described in the ISO standards. That can lead to compliance without capability—a box ticked but no transformation.

The Core Principles

Think of these as the four pillars that keep the whole structure standing:

- **Value.** Assets exist to create value for the organization and its stakeholders. Maintenance should therefore focus on maximizing value, not just minimizing cost. For a utility, that might mean reliability for customers; for a manufacturer, consistent quality and throughput. Value is the reason assets exist.
- **Alignment.** That value only exists through alignment. Asset management decisions must align with organizational objectives. Your maintenance plans should support business goals, not operate in isolation. That alignment doesn't happen by accident—it's built through shared language and priorities.
- **Leadership and assurance.** Leadership must own the asset management system, ensuring capability and compliance. Maintenance excellence can't survive without executive sponsorship. I've *never* seen a world-class maintenance program without an engaged executive behind it.
- **Life cycle approach.** Every asset has a life, from conception to retirement. Managing the entire life cycle optimizes value and mitigates risk. Too

often, we focus only on the "middle years"—operation and maintenance, forgetting that design decisions can lock in 80 percent or more of total cost. The DOE study[4] on maintenance operations highlights this point.

Together, these principles transform maintenance from a cost function into a strategic discipline.

Maturity and Integration

Few organizations start with ISO 55000 certification, and that's fine. What matters first is maturity, not certification. Maturity is like fitness; you don't achieve it once, you maintain it daily.

Ask yourself: "Can we demonstrate, with evidence, that our maintenance, operations, and finance teams make decisions with a shared understanding of risk and value?" If the answer is "Not yet," you know where to focus. When that alignment exists, you're managing assets, not just maintaining them.

Integrating frameworks like ISO 5500x with others—such as ISO 9001 (Quality), ISO 14001 (Environment), ISO 31000 (Risk Management), or ISO 45001 (Safety)—creates a unified management system. It reduces duplication and strengthens governance. Many world-class organizations use an integrated management system (IMS)[5] to streamline policy, assurance, and reporting—and to speak a single organizational language.

You can think of an IMS as a single operating system for your organization: different apps, one platform.

Governance and Policy

Every organization that is serious about asset management needs an asset management policy. It's not just a statement for auditors; it's a declaration of intent, your organizational oath of responsibility for your assets, and an acceptance of accountability for consequences if done poorly. It signals to everyone—from board to toolbox—that asset care is strategic.

A good policy answers:

- Why we manage assets
- What principles guide our decisions
- Who is accountable
- How we'll measure success

Governance extends beyond policies to roles, responsibilities, and decision-making authority. The board must set direction, executives must ensure resources and oversight, and managers must execute within clear boundaries. *When those roles blur, chaos returns.*

When governance works, maintenance teams stop chasing emergencies and start making informed trade-offs among cost, risk, and performance—just as ISO 55000 envisions and as described throughout *Steadfast.*

The Business Case for Standards

Some executives balk at the idea of certification—it sounds bureaucratic. But organizations that embrace ISO 55000 principles (with or without formal certification) often find tangible benefits:

- Reduced unplanned downtime and maintenance costs
- Improved investment decision-making
- Stronger regulatory compliance
- Better stakeholder confidence
- Clear accountability and communication

It's not about ticking boxes; it's about running your business with asset consciousness. In other words, ISO 55000 doesn't slow you down—*it clarifies the path.*

Putting It into Practice

How do you bring this to life?

1. **Start with assessment.** Use a maturity model or gap analysis to compare where you are versus ISO 55001 expectations.

2. **Build an asset management strategy.** Define goals, quantifiable objectives, decision frameworks, and performance measures.
3. **Integrate with maintenance and operations.** Make sure maintenance planning, reliability engineering, and supply chain all align with the asset strategy.
4. **Engage leadership.** Ensure the executive team understands its role in governance, risk, and assurance.
5. **Communicate and educate.** Bring maintainers and managers into the process; standards succeed only when people see their relevance.

I've found this five-step approach works, regardless of sector and industry. And it starts with intent, not paperwork

The Road Ahead

As digital transformation rewrites how we work, ISO 55000 remains the compass that keeps us oriented toward value.

In the end, frameworks and standards aren't the goal. They are the *scaffolding* that supports continuous improvement and sustainable excellence. The real goal remains what it's always been: sustained high asset performance.

That's what keeps us grounded, no matter how technology evolves.

CASE STUDY

Turning Maintenance into Strategy—the Middle Mountain Utilities Story

At Middle Mountain Utilities,[6] *the maintenance shop was full of good people and bad habits.* They saw the ISO 55000 journey as "just another compliance project." The executives saw it as "something for the engineers." Neither group realized it would transform how the entire company thought about value and risk.

The Situation

Middle Mountain operated an aging fleet of water-treatment facilities. Equipment reliability was declining, maintenance costs were climbing, and regulators were scrutinizing environmental performance. The maintenance crews were heroic: working nights, patching leaks, and keeping pumps running. The business was in "firefighting" mode with a *break-then-fix* approach to maintenance. They didn't plan, schedule, or even get most of their PMs completed. "While we were proud of our people, we were burning them out," recalled the COO.

One executive finally asked a couple of pivotal questions:

"If our assets are the heart of our business, why don't we look after them better? Why are we managing them like they are liabilities?"

That question started their asset management transformation.

The Approach

Instead of jumping straight to certification, Middle Mountain started with a gap assessment against ISO 55001. The review revealed what everyone suspected but had never articulated:

- No formal asset-management policy
- Fragmented maintenance and finance data
- Capital decisions made without life-cycle cost analysis
- Confusion over accountability among operations, engineering, and supply chain

The assessment gave shape to their frustrations by showing *it was the system, not the people*, that was failing.

They formed a cross-functional asset management steering committee comprising operations, maintenance, finance, risk, internal audit, and IT to build a shared understanding of value. Together they developed:

- An asset management policy linked to corporate strategy
- An asset management strategy defining how performance, cost, and risk would be balanced
- A governance model clarifying decision rights—from the board to frontline supervisors
- Integrated KPIs connecting maintenance metrics (availability, PM compliance) with business outcomes (cost per cubic meter delivered, customer reliability index)

These became the backbone of their asset-management system.

That also led to the development of various plans, both for the assets and for transition of the organization's behaviors and some challenging implementation efforts. They discovered that culture change, not process design, was the hard part.

The Results

Within 18 months, the change was dramatic:

- Unplanned maintenance reduced by 30 percent.
- Maintenance backlog dropped, and so did total spend, because the right work was being done.
- Capital investment proposals began to include life-cycle cost justifications, improving ROI.
- Executive reports now featured asset health indicators alongside financial performance.
- Perhaps the most telling result was that morale had improved at all levels—the real payoff was cultural.

When ISO 55000 auditors eventually visited, certification came almost as an afterthought. Maintenance and the business were finally on the same page. That alignment changed conversations—and careers.

The Lesson

Middle Mountain discovered that ISO 55000 wasn't about paperwork, it was about alignment. The framework gave structure to conversations that had never happened before: between maintainers and accountants, among reliability engineers, operators, and the CFO.

As one maintenance manager put it:

"We used to justify our work in terms of wrench hours. Now we justify it in terms of customer value."

That's what asset management excellence looks like: maintenance as a strategic business partner, not just a support function.

That's the power of a framework; it turns intent into shared practice.

Key Takeaways

Frameworks bring alignment.

ISO 55000 and similar standards don't replace good maintenance; they give it context. They connect the reliability engineer's toolbox to the executive's dashboard, ensuring everyone pursues the same objective: *maximizing asset value.* That's the sweet spot where *Steadfast* performance lives.

- **Asset management is a leadership discipline.** Maintenance excellence depends on leadership, governance, and clarity of purpose. Without visible executive ownership, even the best maintenance practices plateau. With it, they thrive.
- **Value depends on a balance of performance, risk, and cost.** True asset management is a balancing act. ISO 55000 reminds us that cost reduction alone is not the goal; sustainable performance and risk control are.
- **Maturity comes before certification.** You don't need a certificate to start acting strategically. Use the principles to assess your current state, close gaps, and build cross-functional understanding. If desired, and if it makes good business sense, certification can follow naturally.
- **Governance links policy to practice.** Asset management policy and strategy are only as strong as the governance behind them. Decision rights, accountability, and measurement must be crystal clear, from the boardroom to the toolbox.
- **Culture is the ultimate differentiator.** Standards provide structure; culture brings them to life. Organizations like Middle Mountain succeed because they engage everyone: maintainers, managers, and executives, in a shared vision and journey leading to reliability and value creation.
- **Frameworks future-proof the business.** As the digital era reshapes how we collect and interpret asset data, ISO 55000 offers stability. It can be your compass that ensures new technologies serve strategic intent, not the other way around.

With the right frameworks in place, your organization has structure, governance, and shared purpose. You can help amplify that through the capability of technology: leveraging data, connectivity, artificial intelligence, and analytics to make asset management more predictive, intelligent, and value-driven.

That's what I call "conscious asset management." It's how maintenance grows up into strategy.

CHAPTER 13

DIGITAL TRANSFORMATION AND THE FUTURE OF MAINTENANCE

I don't use the term "data-driven"—we are the ones driving; the data and information it provides in the right context enable us to make better decisions. It doesn't tell us what to do.

The Digital Wave Has Arrived

Walk through any industrial site today, and you'll see signs of the digital shift: sensors blinking, network nodes gathering wireless data from machinery, tablets replacing paper work orders, and dashboards on screens above shops and in control rooms. What used to be manual, largely reactive, and heavily reliant on experience is becoming connected, more predictive, and data-enabled.

Many of the early digital transformation initiatives failed to live up to the hype that preceded them. The software and hardware producers were keen proponents and promised much. But the promises of results were hollow. Results require a different set of behaviors, not just the rollout of new technologies.

Digital transformation isn't just about new tools; it's a new way of thinking about reliability: how we collect, interpret, and act on information about our assets.

The challenge isn't adopting technology; it's using technology to create business value. That's the difference between "digital activity," what many have done, and "digital transformation," what the visionaries want, and eventually, what we will need to remain competitive.

From Data to Decisions

Data is just numbers: ones and zeros with no meaning unless it comes in context. That context must be meaningful to decision making, otherwise it is little more than distraction. We want to be informed, not distracted.

Despite decades of digital tools—from computerized maintenance management systems (CMMS) to spreadsheets—performance hasn't improved much. Information still lives in silos, quality remains poor, and most of all, users don't trust the data.

Today we have hundreds of options when it comes to software. The market size for such systems is estimated to be nearly $1.5 billion. There is much marketing hype, many promises, yet that proliferation of software and availability of data hasn't actually improved performance by very much. Sometimes it even takes us backward!

The problem was, and still is, the fragmentation of the information and data. Data lives in business silos. Some data is in the CMMS, some in spreadsheets, some in people's heads. CMMS systems are notorious for having data missing or incomplete, misleading, and not particularly useful for business decision making. Many users simply don't trust it. Spreadsheets are widely used to manipulate data because the systems designed to do it often don't—or don't do it well. In doing reliability analysis work with many clients, I've observed that the best data about failures is invariably in the heads of the technicians, not the CMMS and not the spreadsheets.

Digital transformation aims to connect all that information into a single view of asset health. With sensors, industrial internet of things (IIoT) devices, and cloud analytics, we can now see what's happening in real time and often predict what will happen next. Digital twins excel those predictions and are great tools for "what if" analysis.

But more data isn't automatically better. Insight is the goal—turning raw data into decisions you can rely on. That requires three things:

1. **Data quality.** Bad data is still bad, even if you digitize it.
2. **Integration.** Connect systems so that planners, engineers, and executives see the same truth.
3. **Context.** Understand what the numbers mean for operations, risk, and value.

Without these three—quality, integration, and context—data is just noise in digital disguise. As one vice president of operations said, "We don't need another dashboard. We need answers."

Proactive Maintenance

Proactive maintenance is about dealing with the consequences of equipment failure to make them tolerable to the business. In some cases, that means preventing them altogether, in others it means finding them in their infancy so actions can reduce the consequences, and in others it is about finding them failed before they are needed to act.

Preventive

In the old days, preventive maintenance was king. We changed oil and replaced parts on schedule, not because they'd failed, but because they will eventually. But not everything fails at a known and consistent frequency. In fact, most failures are far more random.

Predictive

Then came predictive (condition-based) maintenance: Find defects in early stages of failure and fix them before they break. Use the warning time to minimize the consequences of those failures.

Predictive maintenance (PdM) is now using edge computing in IIoT devices and data analytics, sensors, and machine learning to forecast failures before they occur.

Detective

Detective maintenance focuses on discovering hidden failures in protective systems—those that sit dormant until needed. We have an increasing reliance on safety devices and backup equipment to maximize reliable performance. If they fail, disaster can result.[1] We now realize the need to test them so they are available to work when needed.

Prescriptive

The concept of prescriptive maintenance has arisen where the predictive maintenance goes a step further with the computer (specifically agentic AI) now recommending actions, balancing cost, risk, and performance. My concern with this is the replacement of human intelligence about system operational state, the condition of equipment and systems nearby, the health of the sensors themselves, and just what artificial intelligence (AI) is doing in the background. The AI agent can get it wrong and can be biased, producing erroneous outputs. It also relies on GenAI in making predictions. There are already many cases of GenAI being too "creative" and prone to hallucinating in what it produces. Nevertheless, technology is incredibly helpful, and leveraging it is only in its early stages.

How Technology Fits

Here's an example of what happens with predictive maintenance showing where technology can help us:

- A vibration sensor detects early bearing wear and sends the alert with data to your vibration analysis software.
- The software uses a rules-based AI model to determine just what is wrong and produces a forecast failure within 28 days.
- With predictive maintenance in place, the system could jump to creating the work order, but few of us are ready to trust decisions normally made by engineers to a computer algorithm programmed by people who don't know much about maintenance.

- Your maintenance engineer reviews it and confirms it's a valid "catch" and action is needed. The engineer may have made a field trip to confirm.
- Using the vibration software, the engineer sends a notification to your CMMS, which creates a work request.
- Your planner attaches the job plan they created with an AI planning tool to the work order. Job status is now "planned, awaiting parts."
- Your materials coordinator confirms materials from the plan are in stock or arriving soon. Once confirmed, the work order is "ready to schedule."
- The scheduler (or your planner) puts the job into the next week's schedule based on its urgency (priority) and the importance of the equipment to the production process. The job is now "scheduled," while parts are being kitted in the storeroom and made ready for pickup. Once everything is there, the job is "ready for execution."
- Production knows the schedule and makes the pump ready for repair by isolating it from the process, draining it, and using another pump to take over its duty. Production won't be lost.
- The assigned trades get the work order, proceed to stores, get the parts kits, and do the job before the equipment has a chance to fail.

The possible breakdown was avoided, the job was done in the weekly schedule, and overtime and rush orders were avoided. This actually happens today in proactive organizations with or without the technology. The technology is built into the process, not the other way around. It adds cost and complexity to your IT (which may not be justifiable), makes it a bit easier to diagnose the problem, and automates parts of your work management process.

The key takeaway: Even the smartest systems can't replace a sound work management process built on discipline and human judgment.

The Role of AI, Digital Twins, and Analytics

AI is rapidly becoming the new technician's assistant and the manager's advisor. There are three broad categories of AI: rules-based, agentic, and generative.

Rules-based AI follows preprogrammed logic: If this, then that. For example, if the vibration sensor picks up a high level at running speed, you have imbalance.

Rules-based (conventional) AI algorithms can process millions of data points from equipment sensors, identify patterns invisible to the human eye, and provide early warnings. As it does this, it also learns. If it's told it got the warning wrong, it learns what not to predict. If it knows it got it right, it will reinforce the algorithm.

Agentic AI learns to be the "agent" that replaces someone. For example, a chatbot answering your questions is powered by an agentic AI following a combination of script and redirects depending on your inputs or questions.

Agentic AI is intended for autonomous decision making like described for prescriptive maintenance but relies somewhat on GenAI and can be prone to mistakes and bias. Bias can be learned or built in by its programmers. Bear in mind that programmers aren't reliability engineers or specialists in the technology (equipment and systems) you want to use it for. It does have a lot of promise and is still in early stages of adoption.

Generative AI (GenAI) doesn't rely on rules. It reads a lot of data from numerous sources and recognizes patterns, then makes creative predictions using what it has learned. For example, it can learn about predictive maintenance and recommend what might work for your plant if you ask the questions the right way.

GenAI is sometimes a bit too creative and is known to "hallucinate"; yes, it makes stuff up. For best results, you need to know how to ask it questions and then carefully review what it produces. Lawyers' cases have been thrown out by judges who recognized the cases being cited were actually invented by an AI and not real. Students' papers are being failed by professors who recognize flaws that the AI created and the student missed.

There are some legitimate security concerns with the use of AI that IT departments are only just beginning to understand, such as:

- Bias in decision making
- Cloned voices and "knowledge bases" masquerading and misleading
- Violations of data privacy through the use of pirated or unreliable data sources
- Its inherent openness to the internet means no air gaps, rendering it unsuitable in high-security environments
- You don't control, or even know, just where your data is being stored, nor who has access to it. That would be a big concern if you are normally handling private, sensitive, or even classified information.

The use of AI is also highly energy-intensive with its high demand for computational power.[2] It may be incompatible with organizations that are looking to be "greener."

In some industries where validation is important, like pharmaceuticals and food, you will be unlikely to explain how decisions are arrived at using AI and risk validation.

It can infringe on intellectual property rights, can be used to replace people (job losses), and can't take responsibility nor be accountable for errors that may result in substantial losses. And when something goes wrong, keep in mind that you can't choke the throat of a computer—but as an executive you will be held accountable regardless.

That's why leaders must understand not just what AI can do, but what it should not do.

Digital Twins

Digital twins are virtual replicas of physical assets. They simulate real-world performance, allowing engineers to test scenarios before making real changes in design or configuration of systems. It reduces the risks that would otherwise come with testing something new in real plants or live systems.

Analytics

Analytics platforms bring it all together: integrating maintenance, operations, and finance data into one coherent decision space.

A partner analytics firm can use the power in data coming from multiple, nonintegrated systems to identify a number of areas for improvement. PMs that are causing problems are identified by breakdowns that follow them closely. PMs that work, but aren't being done often enough, can show up occassionally, but not always followed by breakdowns. A quick optimization of those PMs can reduce work and breakdown. They find plenty of parts that are being direct ordered that should really be in stores by virtue of the quantities being consumed. Put them in stores, improve stores' availability, and reduce your direct (usually rush) purchasing costs. By comparing what is used over a suitably long time and considering the equipment those items are supposed to be used in, they can also identify excessive inventory. And interestingly, they also find parts from direct orders that were never reported as used in the work for which they were purchased. When comparing that list to parts they can physically find in cabinets and other hiding spots in shops and work areas, they realize that they are holding a reasonably accurate listing of their shadow inventory. Those items belong in stores, not your shops.

Integration through analytics delivers measurable savings!

Technology enables smarter maintenance and faster, better decisions. It links field data to action and ultimately, reliability to profit.

Cybersecurity and Data Integrity

As digitization grows and our assets get smarter, so must our defenses. A connected plant can be efficient; but also exposed.

In addition to the concerns arising from AI described previously, cybersecurity is becoming a big deal. It can no longer remain the IT department's concern alone. Maintenance teams must ensure that connected assets—sensors, gateways, control systems—are secure and properly maintained.

Real-World Consequences of Weak Cybersecurity

In one recent case, a manufacturing facility was shut down because its operating technology had been hacked and there was no backup. A major utility was hacked, exposing customer data and taking months to recover. While recovering, most of their projects and initiatives were put on hold!

A compromised device can disrupt production just as easily as a failed bearing. That's how Iran's uranium centrifuges were intentionally harmed by the Stuxnet virus, the world's first cyber weapon, in 2010.

Equally important is data integrity. Without trust in your data, you can't trust your decisions.

Digital transformation succeeds only when the organization treats data as a strategic asset.

The Human Side of Digital Change

Technology changes in months; people take years.

Indeed, technology has already left many of us in the digital dust. If users can't use it correctly, it doesn't help—it becomes a risk or even a problem.

One of the biggest mistakes organizations make is assuming digital transformation is a technical project. It's not—it's a cultural transformation. Those are never easy and if led by technical people, they usually fail.

Frontline maintainers may worry that AI or automation will replace them. I already see that with an AI-enabled planning tool that raised exactly that concern. Planners feared it would replace them. In truth, it made them faster and added safety insights they had been missing. It elevated their role. It won't replace them, but in all honesty, a planner using it could.

Managers, too, must evolve from supervising work to interpreting insights and guiding decisions. Executives must champion this change, investing not only in systems but in skills and mindset.

As one plant manager said after implementing predictive analytics in their production processes: "The hardest part wasn't installing sensors and getting it working, it was changing how we validate and trust data, then what we do with what it tells us."

Training, communication, and inclusion are key.

People will support what they help create, but they will balk at being changed.

The Strategic Lens

Digital transformation is not an IT upgrade—it's a business reinvention.

For executives, digital transformation is not a tech expense—it's an investment.

It can be used to strengthen competitiveness, resilience, and agility, but only if you use it strategically and not as a technical implementation of a new tactical tool.

To deliver on the lofty tech promises, those tech initiatives must tie directly to the business strategy.

Executives should regularly ask three questions to ensure digital efforts stay strategic:

1. Does our digital roadmap align with our asset management objectives?
2. Are we prioritizing technologies that solve real business problems?
3. How will we measure value: through availability, production, cost, risk, or sustainability?

Digital tools should not exist in isolation; they should enable the ISO 55000 principles: generating value while balancing performance, cost, and risk.

The most successful organizations treat digital transformation as a continuation of their reliability journey, not some separate project in, and led by, an IT department. Digital transformation extends the same disciplined mindset that underpins precision maintenance and reliability.

A Digital Warning

Technology is a powerful enabler but not a substitute for leadership, discipline, or accountability. Many major failures stem not from a lack of technology, but from over-reliance on it. When systems are treated as fail-safe rather than fail-soft, complacency sets in: warning signs are missed, skills erode, and risks go unchallenged until they surface catastrophically. In a Steadfast context, technology must reinforce standards not replace them and data must inform decisions not make them. Precision still depends on capable, engaged people, with human judgment as the final line of defense.

A Glimpse into the Future

Looking ahead, the boundaries among operations, maintenance, and business intelligence will continue to blur.

We're entering an era of ecosystems that sense, decide, and act—with or without us:

- **Fully connected ecosystems.** Assets that talk to each other and to people seamlessly
- **Autonomous maintenance.** Systems that schedule and execute minor tasks automatically. Some of this already exists!

- **AI-driven decision support.** Where leadership decisions are guided by predictive insights. Even if you don't realize or authorize it, don't be surprised if your engineers aren't already doing this.
- **Sustainability integration.** Maintenance practices optimizing not just cost and performance, but also carbon footprint and resource efficiency. Let's not forget that U.S. DOE report[3] linking 20 percent savings in energy by eliminating parasitic losses through precision maintenance.

The future belongs to organizations that can combine technology with discipline, data with judgment, and automation with human wisdom.

CASE STUDY

The Smart Refinery: "Zero Surprises"

At a major refining complex in Southeast Asia, reliability had always been a battle. They had thousands of rotating assets, hundreds of skilled technicians, and seemingly endless unplanned downtime. Availability was well below industry benchmarks, and only low labor costs were keeping them afloat—an advantage that was rapidly disappearing.

When they launched their digital transformation program, they didn't start with sensors, they started with a vision: "Zero surprises."

Over three years, they:

- Deployed IIoT smart condition monitoring sensors across critical equipment
- Integrated their CMMS, ERP, BI, and process control systems
- Began using the AI-enabled software from their condition monitoring suppliers to predict failures, in some cases up to 45 days in advance
- Tied it all together with a central reliability "war room" with live dashboards showing what was happening (plant status), what they were finding (the faults detected), actions taken on those findings, measured response and resolution times, and results (availability and plant output)

The results were stunning: downtime reduced by 40 percent, maintenance costs dropped 20 percent, and safety incidents declined sharply. Remember the link between safety and reliability!

What led to the change? Their executives had a vision of maintenance as a strategic enabler of operational excellence to achieve "zero surprises"!

Their success came not from technology alone, but from leadership's clear vision: reliability as strategic advantage.

Key Takeaways

1. Start with value, not tech. Digital transformation is about value and culture, not technology. Tools matter only if they improve decision quality and business outcomes.
2. Predictive becomes prescriptive. Eventually, prescriptive maintenance will extend reliability's reach, transforming maintenance from reactive and cost-focused to proactive and strategic advantage.
3. Data integrity and cybersecurity are foundational. Without trust and protection, digital systems will be at high risk of failure to deliver value.
4. People drive transformation. Skills, confidence, and culture need attention before you can realistically expect results from machines.
5. Digital tools such as AI and analytics help bridge maintenance data into asset management. They operationalize asset management principles through data, analytics, and intelligence.

Looking Ahead

Digital transformation is part of the story, but people are a much bigger part. Without them, the computers can talk to themselves and to each other, but nothing really changes in the field.

Technology enables change; leadership initiates and sustains it.

In the end, it's not AI that transforms maintenance—it's human intelligence, amplified by the right tools.

In the next chapter, we'll explore how to lead the cultural shift, because the ultimate power is human, not digital.

CHAPTER 14

LEADERSHIP, CULTURE, AND CHANGE MANAGEMENT

Peter Drucker is often credited with saying that "culture eats strategy for breakfast," a reminder that shared values and behaviors drive success more than any strategy.

That insight plays out repeatedly in maintenance and reliability settings; I've seen that in action with the introduction of reliability-centered maintenance (RCM), planner training, redesigned business processes, and more.

Shifting Culture

Even the best systems and tools fail without the right culture. In maintenance and reliability, culture determines whether the organization's efforts toward improvement will take root or be rejected as another passing initiative. A reliability-focused culture doesn't happen by accident; it is shaped over time through leadership, consistency, and example.

A major utility introduced RCM, eliminated a lot of ineffective PM work, and immediately ran into union resistance. People didn't understand exactly what was going on, why, or how the decisions were being made. That misunderstanding reflected a deeper issue: People didn't grasp the purpose or process of change.

Cultural change begins when people understand why a shift is necessary. Leaders must help teams see the connection between reliability and organizational success: lower costs, safer operations, greater asset availability, and pride in craftsmanship. They must also show how the changes will matter to the employees. They won't all care about costs and availability, but they will care about pride, mastery, and meaning.

Belief alone is insufficient; real cultural change requires removing contradictions between stated values and actual practices. For example, if management claims to value proactive maintenance but continually rewards short-term production gains at the expense of long-term reliability, credibility collapses. The culture reverts to expediency.

Successful cultural transformation often begins with a burning platform or, just as often, an opportunity that is being missed—a recognized need to change—and progresses through reinforcing experiences that prove the new way works; each success story, no matter how small, becomes part of the new narrative. Leaders should amplify those stories, using them to model desired behaviors and embed reliability as a shared organizational identity.

In the utility RCM example, the union became suspicious that jobs would be lost. By training union members in the RCM process, they gained understanding and saw how it enabled them to do a better job , and the pure logic of RCM trumped the fears they had about management intentions. Resistance turned into support, and soon they were active participants.

Building a Reliability-Centered Culture

A reliability-centered culture is one where everyone—operators, maintainers, engineers, and managers—understands their role in sustaining reliable operations. It doesn't belong to maintenance alone. Reliability becomes a shared value that cuts across silos, aligning the organization's goals, systems, and daily actions.

The foundation of this culture rests on three pillars:

1. **Shared understanding.** Employees grasp the principles of asset reliability, from condition monitoring to precision maintenance. They know why these practices matter and how they contribute to safety, performance, and profitability. Education and communication are critical here; peo-

ple support what they understand. As the utility case showed, once people understood the rationale, they supported the change.

2. **Empowered teams.** Operators and maintainers are trusted to make informed decisions. They have access to accurate data and the authority to act on it. Empowerment requires psychological safety: the freedom to raise concerns, report failures, or suggest improvements without fear of reprisal. In that same example, it was the RCM analysis teams—not management—who made the decisions.
3. **Aligned systems and incentives.** Performance measures, recognition systems, and management processes must all reinforce reliability objectives. When incentives reward reactive behavior or short-term output, they inadvertently undermine reliability culture. Aligning KPIs with long-term asset performance fosters consistency between intent and action. Although KPIs and financial rewards didn't change, the ability to make a real difference in service delivery appealed to the professionalism of the utility's tradespeople.

Building such a culture demands visible leadership engagement. Leaders must consistently demonstrate reliability behaviors, asking questions like, "What are we doing to make sure that doesn't happen again?" not "What went wrong?" One is positive and forward-looking; the other, negative and backward-looking. They should be involved in reviewing recommendations from the field, participating in improvement discussions, and celebrating proactive wins. In reliability-centered organizations, leadership presence on the shop floor is not ceremonial—it's an act of commitment.

In a large mining operation in southern Africa, the managing director was actively and visibly engaged. He participated in early training sessions; asked trades, planners, and supervisors what needed improvement; and thanked them for their input. He shared what he learned with the leadership team and followed up on it. Over time, employees who once feared his visits began greeting him openly, sharing ideas freely. He always listened and acted. That transformation was worth over $2 billion in additional revenue per year after three years of sustained change. His participation helped them gain $800 million of that in just the first year.

Sustaining Excellence Through Leadership Commitment

To engineers, change is an event; to everyone else, it's a process.

Change is not a one-time event; it's a sustained evolution. The most difficult challenge in reliability improvement is not achieving initial success, but keeping it alive once the novelty fades and the attention is elsewhere. This is where leadership commitment defines long-term excellence.

Sustained reliability requires leaders who act as stewards of culture. They reinforce the message that reliability is integral to the organization's identity, not a project with an end date. They also ensure that management systems—planning, budgeting, performance evaluation—continue to support reliability goals, even as personnel or market conditions change. Without this continuity, even the best initiatives eventually erode.

Make no mistake about it—this is a leadership function, not a technical one. Many technically oriented managers struggle to serve as stewards of culture because it demands empathy and engagement beyond analysis. Similarly, HR-driven approaches often rely on administrative methods that lack the empathy and engagement required.

Leadership commitment is expressed through:

- **Consistency.** Staying the course, even when short-term pressures tempt shortcuts.
- **Visibility.** Being present, interested, and engaged with operations, maintenance, and reliability teams.
- **Accountability.** Holding themselves and others responsible for actions and changing behaviors that uphold reliability values. They must also be accountable for results.
- **Adaptability.** Recognizing that as technology and workforce dynamics evolve, so must leadership approaches.

Years ago, a command-and-control style worked with baby-boomers. Today's workforce, however, expects to understand and question decisions. For best results, make the transformation theirs. Lead through "command and nurture": set direction, let them decide how, and then support their actions.

Ultimately, sustaining excellence in reliability is about institutionalizing continuous improvement. The most mature organizations embed reliability into their strategic DNA. They measure it, talk about it, and make it part of everyday decisions. Over time, reliability ceases to be a program—it becomes the culture.

Summary

Culture is the soil in which reliability practices either flourish or fail. Shifting it requires vision, patience, and above all, leadership. The organizations that succeed are those where leaders model the behaviors they seek, align incentives with reliability outcomes, and commit for the long haul. When leadership and culture align, reliability excellence becomes self-sustaining—a natural part of how work gets done.

CHAPTER 15

THE NEXT FRONTIER—MAINTENANCE PARTNERSHIPS AND SERVICE MODELS

In the past, we managed maintenance. Now and in the future, we must manage reliability.

A New Horizon for Maintenance and Reliability

Maintenance has come a long way. Once a reactive craft, it became a disciplined management function and is now becoming a driver of operational performance. In recent decades, maintenance has matured into a strategic business capability—a lever for value creation, risk control, and sustainability.

Now another frontier is emerging: organizational transformation. Around the world, companies are beginning to ask whether they must *own* every capability required for maintenance excellence—or whether they can *partner in* reliability.

Contracting and outsourcing are not the same! Contracting involves hiring an individual or company for a specific, often temporary, project with direct management by the owner. Outsourcing entails the delegation of an entire function or department to a third-party company for an ongoing period. Outsourcing may be done to reduce costs, and often is, but it can also be done to bring on capability, expertise, and efficiencies that the owner either can't or doesn't want to develop. Contracting focuses on a specific task or skill, while outsourcing handles broader business operations and often involves less day-to-day control on the part of the owner.

Other business functions have already crossed this threshold. Information technology evolved from internal departments to managed services and cloud platforms. Logistics became third-party ecosystems (3PLs) that deliver entire supply chains as a service. Even HR—once a closely held internal function—now blends in-house leadership with outsourced capability. Each of these transitions began cautiously: not by outsourcing core work, but by rethinking what "core" truly meant.

Maintenance and reliability are next, and this time, the change will redefine how we think about delivery capability.

The idea of outsourcing isn't new—contracting and outsourcing have existed for decades—but the nature of the opportunity has changed. Contracting in maintenance has long been used to supplement the workforce in numbers and in capabilities. For instance, the use of maintenance contractors has long been used to add hundreds (sometimes thousands) of skilled trades for relatively short-duration but high-intensity shutdowns and turnarounds.

Outsourcing became popular in the late 1980s and 1990s. Behind it was a drive for cost savings through the strategic use of outside resources. Anything not "unique" (or "core" as some call it) to the company's business could be outsourced: accounting, HR, data processing. It worked and grew into other areas such as manufacturing, information technology, marketing, and customer service.

In some parts of the world, it expanded more than others. In Europe, the Middle East, Asia, and to some extent Australia, we already have outsourced maintenance. In many of these markets, cultural comfort with shared risk made partnership easier to adopt.

North American companies have been slower to adapt though. Production is usually considered too important to risk with outsiders. That logic, which ties business performance to risk, extends to maintenance because that's what keeps things running. As skilled trades shortages and the need for ever increasing sophistication in how we do maintenance grow, the appetite of companies to deliver needed training and to do succession planning to avoid labor shortages has shrunk. Many companies already use outside contractors for their most challenging maintenance work—why not extend that thinking to encompass all the work?

Traditional procurement processes often lack the flexibility or governance sophistication required for partnership-based agreements. The enterprise management systems that many purchasers rely on are not well designed to handle sophisticated agreements as opposed to line-by-line purchase orders. As a consultant, I am often confronted with dozens of pages of "standard terms" that are irrelevant to the sort of work being done, and less than a page to describe the desired work scope. When purchase orders are made up of various line items as if needed to list parts you are buying, it is difficult to describe an outcome that is performance related.

Here in *Steadfast* we are shifting thinking from maintenance as a "service to operations" to "reliability as a strategic advantage." This shift challenges us to ask a tough question about our existing operations: If reliability does provide such a strategic advantage, why have we not yet achieved it sustainably? The answer lies within and the situation you are in has probably been that way for a long time. Should you work to fix that with people who obviously don't know how to do it? Or should you look for someone who is good at it to do it for you?

Executives need to stop asking, "Can we trust outsiders with our equipment?" and start asking, "How can we collaborate to achieve reliability outcomes faster, smarter, and more sustainably?"

The Forces Driving Change

Several converging pressures are reshaping how organizations think about maintenance delivery:

- **Workforce realities.** Skilled maintenance trades are retiring faster than new talent enters the field. Already, the last of the baby-boomers are in their early sixties. As they leave, so too goes a lot of knowledge and experience. Most firms did little or nothing to capture that knowledge before the retirement accelerated during the Covid-19 crisis in 2021 and 2022. Many firms can't afford to maintain a full internal bench of expertise for specialized assets or technologies. Increasingly, they can't find, attract, and retain the expertise, even if they can afford it.

- **Technology complexity.** The digital layer of maintenance—sensors, analytics, artificial intelligence—requires specialized capabilities. These are often better maintained as shared services than as isolated internal systems.
- **Economic and strategic pressures.** Executives want predictable performance and cost outcomes, not endless internal firefighting. Alternative service models promise greater control over availability and total cost of ownership. An outsource service provider may be able to manage your spares, too, freeing up your working capital for investment in what you do best.
- **Risk and accountability.** Modern enterprises are under increasing pressure to manage reliability as part of enterprise risk. It impacts performance in safety, environmental compliance, and even business continuity. Partnerships with reliability specialists can extend risk-sharing beyond the organization's walls. In-house, you'll need a cadre of reliability engineers and quite likely a culture of reliability that isn't there today.

TABLE 15-1 Shifts in Executive Focus

Traditional Focus	Emerging Focus
Managing internal resources	Managing outcomes and performance
Optimizing cost and efficiency	Creating value and resilience
Controlling head count	Leveraging shared expertise
Activity-based contracts	Outcome-based partnerships
Short-term contracts	Long-term performance alliances

Each shift, as shown in Table 15-1, brings opportunity—and new responsibilities for leadership.

From Contractors to Partners

Traditional maintenance outsourcing has been transactional: contractors supply people, time, or repairs, while the owner provides the instructions and takes responsibility for the outcomes. The relationships are mostly commercial, not strategic, especially in larger centers where there is an abundance of competition. However, even in remote locations the same approach to contracting, usually imposed by head offices in those larger centers, is still transactional. The goal is cost containment, not performance transformation.

By contrast, partnership models are built on shared accountability. The external partner isn't just supplying labor—they're co-owning reliability outcomes. As a rule, they also have some "skin in the game" with you—if you win or lose, so do they.

Emerging models include:

- **Partnering for performance (P4P, some call this reliability-as-a-service [RaaS].** Here, performance metrics such as availability and energy efficiency can become contracted deliverables under a service-level agreement. An example of this is aircraft engines: the airline pays for thrust, not an engine.
- **Integrated asset partnerships.** The client retains ownership and governance of the assets, while the partner manages maintenance strategy, execution, and continuous improvement. This enables easy transfer from one partner to another, giving contract managers a bit of flexibility if they feel they need it.
- **Hybrid models.** Internal teams focus on governance and operations coordination, while partners provide advanced analytics, technical specialists, or reliability engineering capability.

The key difference is intent. The purpose of partnership is not to outsource maintenance but to extend capability: to combine the client's operational knowledge with the partner's reliability expertise. If airlines can lease thrust from engine suppliers, why can't you contract for reliable performance and availability?

The goal is not to outsource maintenance—it's to co-own reliability.

What Makes Partnership Work

Partnership requires a maturity of thinking that goes beyond procurement. It succeeds only when capabilities are complementary and when trust and incentives align.

1. **Trust and transparency.** Shared data platforms and open communication are essential. Both parties must see the same truth about performance and risk.

2. **Alignment of incentives.** If one party benefits from more work and the other benefits from less, as it does in most service contracts, then failure is built into the contract. Both must gain from the same performance outcomes: availability, efficiency, and cost-effectiveness.
3. **Capability integration.** Internal and external expertise must be complementary. Operators know the context; partners bring methods, technology, discipline, and scale.
4. **Governance and accountability.** The organization must retain stewardship of its assets. Under ISO 55000 principles,[1] accountability for asset performance cannot be outsourced, but capability can be augmented.

Consider the mining industry, where two distinct models highlight the differences:

Throughout the industry, there are maintenance and repair contracts (MARCs). These are service agreements with well-defined scope limitations in the guise of an outsourcing partnership. These contracts adhere strictly to manufacturer-recommended maintenance, which is often excessive—there is no incentive to the service provider to improve spending on maintenance nor delivery to availability targets. For instance, once your mobile equipment exceeds a fixed number of operating hours, the parts are no longer included, they are extra and now at a much higher price.

In contrast, in the same industry, a contract mine operator is actually more of a partner delivering a target tonnage and taking on full responsibility for operating the mine and maintaining the equipment. If they can't sustain availability, they can't deliver the tons. Deliver more, you earn more; deliver less, you suffer penalties. There is plenty of incentive to get your maintenance program right and operate efficiently. Getting it right requires knowledge, skills, and precision that owners in this industry seldom develop.

IT services were among the first to be outsourced, originally under "data processing" contracts. Without governance, performance often fell short, and many functions were later brought back in-house. They didn't get it right, suffered, and then changed the model.

Governance is important in a partnership; without it, you are not outsourcing responsibility, you are abdicating, and at the end of the day, you still hold all the accountability if things go south.

What About Your People?

I've presented and discussed this concept with large groups of maintainers at conferences. The initial reaction to the idea of outsourcing is "No way!" As a result of the use of outsourcing to lower costs, usually by laying off staff, the word has gained a reputation for meaning "layoffs." But in maintenance, where we already can't find the skilled people we need, that doesn't make sense. Why would anyone lay off resources that are really hard to find, hire, and retain? They are far too valuable. Outsourcing is about adding capability, not shedding costs.

When discussing career possibilities, those same maintainers begin to realize that their careers could actually have much brighter futures. If they are in a business that focuses on maintenance and reliability rather than treating it as a cost center, what they do becomes "core" business. They quickly realize that such a company needs high levels of skills and knowledge, not just brute strength and long hours.

Those maintainers in those discussions shifted from initially negative thoughts to "Where do I sign up?" within less than an hour of discussion. For them, it won't be a hard sell.

A company's skill in managing contracts may need to grow, but that should be relatively easy to achieve with all the available business graduates in the labor market these days.

For executives worried about the reaction of unionized workforces, the job of transitioning will be much easier than they might expect.

Strategic Implications for Executives

For executives, the emergence of service-based reliability models represents a strategic choice, not just an operational one.

- **Shift in metrics.** When availability or energy performance becomes the deliverable, traditional metrics like maintenance cost per unit or technician utilization lose relevance. The conversation turns to value, risk, and resilience.
- **Shift in roles.** Maintenance managers become relationship and performance managers, ensuring the partnership aligns with business priorities.

Technicians may shift from execution to oversight, analysis, and improvement, or may find new opportunities within the partner organization.

- **Shift in risk.** Performance contracting transfers operational risk to the partner but introduces relationship and dependency risk. Leaders must manage these through clarity, governance, and shared success criteria.

The result, when done right, is a more agile organization: one focused on managing outcomes, not inputs. But as with all strategic models, outsourcing present-day dysfunction doesn't work very well. Don't be overly prescriptive about how the partner manages. If your internal processes are chaotic, a prescriptive partnership may only extend the chaos. Partners can amplify good practices—or magnify poor ones. The arrangement with a partner will need to include provision for correcting the chaos and allowance for the flexibility needed to deliver the results you really want.

Glimpsing the Future

Digital transformation can be an engine to help accelerate this shift. As data becomes more transparent, existing inefficiencies will be revealed, and the case for managing through an alternative approach becomes more attractive. We have seen that using analytics to delve into what is happening with proactive maintenance spares management and purchasing in several organizations. When an analytic exercise can reveal big multi-million-dollar savings in just a few days, it becomes easy to see that new perspectives have value. That same capability enables ongoing efficiency, and real-time, even seamless collaboration between clients and service partners.

In this future, reliability performance may become a traded service, measured and priced like cloud computing. The concept isn't far-fetched: Airlines already pay for hours of thrust. Companies might purchase "availability" or "productive capacity" from specialized providers, while focusing their internal resources on innovation, safety, and customer value.

We may see performance ecosystems emerge: networks of organizations sharing tools, data, expertise, and even inventories of spares to manage asset reliability across industries.

Perhaps the next generation of maintenance leaders won't manage a department at all—they'll manage reliability performance across a network of partners.

Reflection

Executives reading this chapter don't need to make an immediate shift, but should recognize the direction of travel. As maintenance continues to evolve from a technical function to a business discipline, its next evolution may be structural: from internal ownership to shared reliability partnerships.

Consider: How would your board measure success if reliability were to become a contracted outcome? If you had a truly specialized expert taking care of it for you, would it be better than it is today?

It's not a question of *whether* this model will mature, but *when*—and who will be ready to lead it. Those who prepare now will define how that future unfolds.

PART IV

INTEGRATION AND ROADMAP TO *STEADFAST* PERFORMANCE

Every improvement journey in maintenance and reliability eventually reaches a stage where technology and methods are no longer the main barriers—alignment is. Aligning people, processes, technology, and business purpose becomes the central challenge. *Steadfast* has shown how the principles of reliability engineering, precision maintenance, and asset management evolve with digital tools, enabling processes run by real people. Part IV focuses on how to bring all of this together across the organization, whether it be a single-site operation or multiple sites spanning the globe.

We begin by exploring how maintainers, managers, and executives can connect their work through a shared "line of sight" from the shop floor to the boardroom. This alignment turns maintenance from a necessary cost into a strategic enabler of performance. It ensures that technical decisions support business priorities and that leadership recognizes the value generated by reliability at every level.

Next, a series of case studies shows what excellence looks like in real organizations: what worked, what failed, and what was learned. These examples highlight that maintenance success is never only about technology or process; it's about people who understand their purpose and act with intent collaboratively across business functions. Culture, leadership, and collaboration make the difference between a good system on paper and sustained performance in practice.

Finally, the "Roadmap to *Steadfast* Performance" provides a structured guide for organizations at different stages of maturity. It outlines a clear progression from reactive to proactive, and ultimately to strategic maintenance, showing what capabilities and practices matter most at each step. The roadmap is practical, but it's also forward-looking: It helps leaders and teams see maintenance as a living system that continuously adapts and improves.

Part IV brings the *Steadfast* framework full circle. It integrates people, process, technology, and purpose into a single direction: building an organization that learns, collaborates, and continually advances toward reliability excellence. The goal is not perfection—it's the capability to improve, sustain, and deliver lasting value from every asset, every day.

CHAPTER 16

ALIGNING MAINTAINERS, MANAGERS, AND EXECUTIVES

Bridging Technical Practice with Business Priorities

Maintenance and reliability form an organized system, and like any man-made system, they are subject to a very natural tendency toward chaos.[1] Put energy into building and sustaining the system and you'll get the performance you want; leave it untended and you'll get chaos. Many are there now, but *Steadfast* ensures you won't be one of them.

No matter how advanced the tools, methods, or analytics become, maintenance and reliability still depend on people, and on how well those people are aligned. When maintainers, engineers, operators, managers, and executives work toward a common purpose, performance improves across every dimension: useful availability, stability, safety, environmental performance, cost, risk, and morale. When that group doesn't work together, those benefits erode and even disintegrate. Even the most sophisticated systems can falter—no organized system is immune.

Achieving high performance with *Steadfast* depends on that alignment, and this chapter explores how to build it. It connects technical practice with business intent, linking the decisions made on the shop floor to the outcomes measured in the boardroom. The goal is a shared line of sight: everyone, at every level, understanding how their work supports the organization's strategy and value creation.

The Line of Sight from Wrench to Boardroom

In most organizations, the people who maintain assets and the people who allocate resources inhabit very different worlds. The maintainer deals with vibration readings, torque values, and safety procedures. The executive deals with balance sheets, production targets, and shareholder expectations. Yet both are managing the same risk: the risk of asset failure and business disruption.

Alignment begins when each group can see through the other's lens. For the maintainer, it means understanding that every technical decision and field action has financial and operational consequences. For the executive, it means recognizing that reliability is not just a maintenance outcome, it's a performance driver.

When a technician takes the extra time to perform precision alignment, they are not just following procedure; they are preserving equipment life, saving energy, reducing downtime, and protecting profit margins. When leadership invests in training, tools, or predictive technologies, they are not just funding maintenance, they are enabling business resilience. The link is continuous and measurable. The key is making it visible.

Translating Between Two Languages

Maintainers and managers often speak different dialects of the same truth. The maintainer talks about mean time between failures (MTBF), root causes, availability, or bearing tolerances. The executive talks about return on assets (ROA), production variability and reliability, and the cost of downtime. Both are describing reliability performance, but without translation, the connection is lost.

Organizations that succeed in reliability excellence build translators into their processes. They develop common metrics and visual dashboards that express maintenance performance in business terms. For example, showing the cost avoided by eliminating a chronic failure communicates far more powerfully than reporting a lower failure rate. Likewise, showing executives how maintenance backlog, resource utilization, and precision work quality influence throughput closes the communication loop.

This translation is not just about language, it's about respect. When each group understands the other's priorities, decisions improve. Maintainers gain context; executives gain insight. Both begin to see reliability as an organizational capability rather than a lofty goal assigned to a costly department.

Leadership at Every Level

Alignment requires leadership from every level, not just from the top. Supervisors, planners, engineers, and technicians each play a role in shaping culture and practice. Leaders in maintenance are those who connect daily actions to long-term outcomes. They encourage disciplined work execution, emphasize continual improvement over fixes, foster learning, and communicate the why behind procedures.

At the management level, leadership means creating systems that sustain alignment. Clear roles, defined workflows, and supportive performance measures all signal that reliability matters. When leaders reward problem-solving and cross-functional collaboration rather than firefighting, the organization shifts from reactive behavior to proactive culture.

Executives, meanwhile, set the tone. Their words and actions determine whether reliability is viewed as a strategic lever or an operational afterthought. The most effective executives link reliability objectives directly to business goals—profitability, customer satisfaction, sustainability, and safety—and hold the organization accountable for results. A good asset management plan or a reliability charter can be used to state the goals, elaborate on the objectives and their metrics, and define just who—or more specifically which team—is responsible for achieving them and who is accountable for the results.

Building the Bridge

Bridging the technical and business worlds is not a one-time effort; it requires a continuous dialogue. The most reliable organizations maintain that dialogue through:

- Shared performance measures that connect maintenance metrics to business outcomes; increased asset availability leads to greater operations utilization and then to productive output.

- Joint reviews of asset health and production performance, involving both engineers and executives; trends in availability, utilization, progress in planned and proactive work.
- Cross-functional teams that include operations, finance, engineering, and maintenance working toward shared KPIs. Each knows how they influence all the KPIs and what needs to happen to improve on them.
- Education and visibility, ensuring decision-makers understand the realities of maintenance work and maintainers understand the business they support.

When everyone can see how their work affects enterprise performance, engagement rises. People take pride in their contribution and understand its impact. The result is not just better maintenance, it's better business.

From Compliance to Commitment

True alignment goes beyond compliance with regulations, procedures, or reporting lines. It reflects a shared commitment to reliability as a value, not just an outcome. In such organizations, a technician's pride in precision, a planner's discipline in planning and scheduling, and an executive's decision to reinvest in reliability all spring from the same belief: that reliable assets enable sustainable success.

This shared commitment defines the culture of *Steadfast* performance. It connects the wrench to the boardroom, turning maintenance from a cost to be controlled into an investment that drives performance, safety, and reputation. The result is an organization that not only runs better but thinks better about reliability.

When every level sees reliability through the same lens, excellence becomes the default, not the exception.

CHAPTER 17

CASE STUDIES IN MAINTENANCE EXCELLENCE

In the preceding chapters, we have the *Steadfast* framework leveraging experience in applying concepts that may seem like theory but actually reflect a lot of common sense and logic. In this chapter, we turn to practice: real-world cases of asset-intensive operations that have moved the needle, and the key takeaways for practitioners looking to replicate such gains. Because *Steadfast* entails a change in culture, process, and technology, the "lessons learned" often reside not solely in the sensor or software layer but in how organizations have made the people–process alignment work.

Here we turn from principle to practice. The following case studies, drawn from both external experience and my own consulting work, illustrate how organizations in diverse sectors have applied *Steadfast* concepts to deliver measurable, sustainable improvement. The most valuable lessons reside not in the technology itself, but in how people, process, and purpose have been aligned.

CASE STUDY A[1]
Transitioning from Reactive to Predictive in Manufacturing

A large automotive-parts manufacturer (operating over 200 critical production assets) adopted a predictive maintenance solution that leveraged IIoT sensors, vibration/temperature/acoustics monitoring, and AI analytics. According to the authors of the case study, the company improved equipment availability from approximately 82 to 95 percent and reduced unplanned downtime from 18 percent of production time to about 4.9 percent.

What They Did

Their path included:

- Conducting a critical-asset inventory and failure-mode baseline (months 1–2)
- Installing IIoT sensors and building analytics models in a pilot on two production lines (months 3–6)
- Rolling out to plantwide deployment, integrating with CMMS and maintenance workflows (months 7–8)
- Prioritizing continuous improvement and predictive-work-order integration

Key Results

Among the claimed outcomes:

- 40 percent improvement in equipment availability (from 82% to 95%)
- 73 percent reduction in unplanned downtime
- ROI achieved in just over a year

Lessons Learned

1. **Start with criticality, not every asset.** The pilot focused on the assets with highest failure impact (computerized numeric control [CNC] machining centers, injection-molding machines, robotics) so the business benefit case was clear.
2. **Integration counts.** The value came not just from sensors or analytics, but from linking predictive insights into the CMMS, work-order prioritization, and execution. Without that linkage, the analytics sits in a silo.

3. **Change management matters.** This shift required technicians and planners to think differently: not "We'll wait until it breaks" but "What is the health score and what are the early warnings?" Training and communications were essential.
4. **Don't chase perfection immediately.** The pilot allowed false-positive tuning, model refinement, and data cleanup before full scale-up.
5. **Metrics beyond availability.** They tracked failure-prediction accuracy, reduction in emergency work orders, parts inventory reduction, and scrap-rate improvement. This aligns well with the U.S. Department of Energy (DOE) Operations and Maintenance (O&M) emphasis on performance measurement.

One might be tempted to look at that manufacturing case study as pure marketing for the IIoT, as I did initially, but the results and observations align perfectly with my own experience as described in case studies D, E, and F as well as others that are not included here.

CASE STUDY B
Precision Maintenance in Facilities Services

Wernher von Braun of the National Aeronautics and Space Agency (NASA) coined the term "precision maintenance" in reference to his findings that a 20 percent reduction in vibrations doubled the useful life of bearings. One recent article[2] titled "Precision Maintenance: What It Is, Why You Need It, and How to Start" emphasizes that precision maintenance is not only about predictive analytics, but about consistent, accurate execution of maintenance procedures, workforce training, and process documentation.

What They Did

The facilities-services provider implemented:

- Standard operating procedures (SOPs) for PM tasks (e.g., lubrication, filter changes, visual inspections)

- A CMMS to track work orders, technician performance, spares usage, and maintenance outcomes
- Training for technicians to ensure consistency in execution (less variability regardless of who does the task)
- Data-collection and trend-analysis for asset health, leading to condition-based intervention

Key Results

While specific numbers were not always published, the article indicates that organizations adopting precision maintenance achieved:

- Reduced downtime of building systems
- Increased asset lifetime
- Improved maintenance cost control
- Enhanced product/service quality (e.g., fewer heating, ventilation, and air-conditioning system failures, better indoor-environment reliability)

Lessons Learned

1. **Procedure standardization is foundational.** If technician execution is inconsistent, then predictive alerts and condition monitoring still suffer—one of the "cultural gaps" the article identified.
2. **Workforce skill matters.** Training and certification of technicians was a core enabler—again aligning with DOE's emphasis on maintenance-personnel competence.
3. **Scalable pilots help.** The article emphasizes starting small in one or two areas, demonstrating value, then scaling.
4. **Don't ignore data hygiene.** For condition-based or predictive analytics to be meaningful, you need reliable data on asset history, maintenance logs, failure records—often a gap in facilities environments.
5. **Link to business outcomes.** In facility services, the connection between maintenance and service-quality (occupant comfort, asset life, regulatory compliance) helps justify investment in precision maintenance.

CASE STUDY C
Digital-Twin and Condition Monitoring in Infrastructure

A recent academic case paper[3] referenced a reinforced-concrete bridge in Norway where a digital-twin model, fed by sensor-data, enabled detection of structural defect and helped identify intervention before catastrophic failure.

What They Did

They combined:

- IIoT sensors embedded in the structure for monitoring strains, displacements, and environmental factors
- A physics-based digital twin of the structure integrated with machine-learning anomaly detection
- Virtual inspection and condition-monitoring dashboards tied to asset-management workflows

Key Insights

- The digital twin allowed not just fault detection but earlier warning of emerging degradation and guided planning of inspections and remedial work.
- The case highlighted the value of combining physics-based models with data-enabled analytics (hybrid modeling).
- Organizationally, responsibility for condition-monitoring, data-interpretation, and inspection planning had to cross civil engineering, operations, and maintenance teams.

Lessons Learned

1. **Condition monitoring at scale requires systems thinking.** In infrastructure, you manage not just machinery but structures, systems, and environmental interdependencies; so precision maintenance here leans toward asset-management integration.

2. **Data is necessary but not sufficient.** The sensor data had to link to decision triggers: When do you inspect, when do you maintain? Without that linkage, you risk data-overload.
3. **Early detection broadens options.** When you detect a fault early, you can schedule non-disruptively rather than emergency shutdown; this drives value in infrastructure environments where downtime is costly or dangerous.
4. **Cross-discipline collaboration is key.** In this case, engineering, inspection, data-analytics, and maintenance workflows had to align—a common theme in precision maintenance implementation.
5. **Asset-life-cycle view matters.** For long-life infrastructure assets, the objective isn't simply longevity but reliability, safety, and life-extension; precision maintenance methods allow that life-cycle perspective to come to the forefront.

CASE STUDY D
Rollout of R&M Standards Globally—Mining Sector

I first became involved with a single mine in a global gold-mining company operating 16 mines in seven countries. We began with a pilot reliability project at one Canadian mine, applying root-cause analysis (RCA), reliability-centered maintenance (RCM), and improved planning and scheduling. The results—lower maintenance costs and higher equipment availability—were so striking that corporate leadership launched a worldwide asset-management program.

Working alongside the parent company's vice president of process improvement and its asset management team leader, we developed a global model for maintenance delivery and a complementary set of maintenance and reliability standards. Representatives from major sites in North America, Chile, South Africa, and Australia cocreated the model, ensuring practical relevance and their own global involvement in deciding on what should be done at all sites.

An assessment protocol was built and used for baseline evaluations across all 16 mines. Each assessment ended with a facilitated workshop where the site team identified opportunities and built its own prioritized roadmap. A year later, after proceeding largely on their own, site reassessments measured progress.

Most sites took ownership and achieved remarkable improvements through commonsense practices (doing what they knew all along was right), training (often with third-party help), and leadership commitment.

Four years later, the company was acquired by a larger miner, and our framework became the foundation for corporate standards still in use today.

Lessons Learned

1. Involve the operating sites in designing the model—ownership breeds commitment.
2. Use assessments to teach, not police.
3. Allow sites to build their own improvement paths; facilitate rather than dictate.

CASE STUDY E
World's Second-Largest Diamond Producer

An African mining company, the world's second-largest diamond producer, operated four sites: two very large and two smaller satellites. Although highly profitable, the mines ran in a reactive "break-then-fix" mode. Recognizing the risk, a new managing director with transformation experience invited me to assist.

A rapid three-week assessment produced more than 80 recommendations per large site. Many overlapped, allowing us to consolidate them into a corporate-level roadmap.

Key elements of the plan:

- A corporate charter with governance for both transformation and steady-state operations
- A steering committee comprising the managing director, site leaders, and myself as external advisor
- Immediate root-cause analyses to resolve high-visibility reliability problems—our "quick wins" that established credibility
- Companywide education in good maintenance and reliability concepts, emphasizing *efficiency* (doing work the right way) and *effectiveness* (doing the right work)

- Site-specific roadmaps developed locally but aligned corporately, supported by experienced reliability coaches

The program ran three years. In the first, additional revenues reached roughly $800 million; by completion, nearly $2 billion in added value was realized. Costs dropped about 20 percent through better work management, reduced overtime, and lower contractor and materials expenses.

Staffing declined only through attrition—no layoffs—as proactive work replaced repair firefighting.

Lessons Learned

1. Combine clear governance with local ownership to ensure alignment without bureaucracy.
2. Secure early wins to build confidence.
3. Pair training with onsite coaching to embed habits.

CASE STUDY F
Upstream Natural Gas Liquids

A midsized Canadian energy company extracted natural gas and produced condensate across a vast, remote region. Field operations ran largely unattended, controlled from regional centers, and responded reactively to shutdowns. Maintenance crews were constantly firefighting, working long hours, and traveling great distances.

Processing plants, by contrast, practiced some preventive maintenance but still suffered from failures that consumed their capacity. The field and plant operations functioned as separate silos with different philosophies.

I was engaged to assess both and deliver training. The assessment exposed unsafe and inefficient practices, while production and downtime data revealed the business cost of reactivity.

During training sessions, field and plant personnel identified their own improvement ideas—many already half-formed but never validated. Once leadership saw the potential, they approved implementation.

Actions and Results

- Introduced proactive work processes and safer practices
- Trained technicians in diagnostic technology and work-management systems
- Supplemented field teams with contractors during transition to handle backlog
- Within six months, reactive work fell sharply, site visits by field maintenance crews became shorter and safer, and overall availability and output climbed steadily

Lessons Learned

1. Breaking down silos unlocks enormous value.
2. Empowering teams to generate their own solutions sustains change.
3. Investing early in simple, usable management systems accelerates improvement.

Cross-Case Synthesis: Six Common Lessons

Across all cases, six consistent principles emerge:

1. **Prioritize critical assets.** Early focus builds momentum. All successful implementations started by identifying the most critical assets (highest risk, highest cost of failure) rather than attempting a "big-bang" across all assets.
2. **Align people, process, and technology.** The triad is inseparable. Tech solutions alone (sensors, AI, CMMS) don't guarantee results. Standardized processes, trained personnel, and change management are equally essential.
3. **Turn data into decisions.** Insight without action has no value. Condition-monitoring or analytics must feed maintenance decisions: when to intervene, what to intervene on, and how to evaluate outcomes. If analytics remain isolated, value is lost.
4. **Adopt iterative improvement.** Pilot, learn, refine, scale. Our experiences suggests that pilot-projects work to prove concepts and gain early buy-in. Refinement of models and processes using lessons learned (pay attention

to the folks in the field) and then scaling only after early wins. The culture of continuous improvement is central.

5. **Measure what matters.** Success is proven in business outcomes. Maintenance investments succeed when tied to tangible outcomes (e.g., reduced downtime, extended equipment life, lower costs, improved quality, better service). The case studies clearly show this.
6. **Strategic governance anchors tactical success.** Across all cases, improvements accelerated when reliability and maintenance were treated as strategic capabilities rather than departmental activities. In every example—from mining to manufacturing to energy—executive-level sponsorship created the mandate for disciplined implementation and insulated teams from short-term pressures. Governance structures such as steering committees, cross-functional working groups, or corporate reliability standards provided the stability needed to sustain change.

Implementation Pitfalls and How to Avoid Them

In addition to the positive lessons, your *Steadfast* implementation journey is likely to encounter predictable pitfalls. Some of these have been teased out in the previous case studies; others are documented in the literature.

- **Overemphasis on technology.** Some organizations invest heavily in sensors and AI before establishing the maintenance-execution foundation. Without process discipline, results will disappoint. Lay the groundwork—asset criticality, process design, practices and standards, technician training—before major spending on analytics. Beware of hardware and software providers and consulting firms that are quick to encourage technology-driven solutions.
- **Data-quality deficits.** Poor or inconsistent data (maintenance history, failure records, sensor calibration) compromise model accuracy. Historical maintenance data is often unfit for purpose, but sometimes analytics can spot problems and compensate for some of it. Audit and clean your data, start with a clean baseline, and manage data governance. If you are embarking on a system upgrade, don't just pull in the old data.

- **Change-resistant culture.** If maintenance technicians view any initiative as replacing them or as "Big Brother," resistance occurs. *Steadfast* performance must be both good for them and seen to be good, not a threat. Engage the workforce early, emphasize that the *Steadfast* principals depend on their knowledge and elevate their role (from reactive firefighter to proactive optimizer). There may well be a financial argument about loss of overtime—get your financing and HR functions on board and don't shy away from offering incentives for achieved results. They'll be far more valuable than the spending you use to lower resistance.
- **Organizational misalignment.** Urgency often overwhelms importance. When short-term pressures or misaligned objectives dominate, they invariably erode strategic reliability discipline and pull the organization back toward reactivity. Planning, precision work, RCM analysis, and the thoughtful deployment of results all require protected time and stable priorities. If everything is treated as an emergency, nothing becomes strategic—and reliability improvements collapse under the weight of reactive behavior.
- **Unclear metrics.** Without clear targets (e.g., percent reactive work, percent planned and scheduled work, schedule compliance, availability, downtime hours, cost per unit of output), you cannot measure success. Define a scorecard aligned with business drivers, and measure before, during, and after implementation. Show results as well as process improvements—link them.
- **Scope creep.** Some organizations try to roll out broadly before stabilizing early successes, which dilutes focus and slows momentum. Maintenance and reliability can get complicated with many moving parts. Pilots work well, but don't drag them on too long. Maintain laser focus on the pilot assets, document wins, then scale.

Aligning *Steadfast* and Asset Management Principles

This chapter's cases reinforce our earlier framing of *Steadfast* and asset management as strategic enablers of high performance. Reductions in life-cycle cost, risk-based planning, performance measurement, and continuous improvement are all aspects that were manifested in the previous case studies and in my experience elsewhere.

Efficiency in maintenance processes and execution, precision maintenance, reliability management, and asset management all fit well together in this broad *Steadfast* framework.

Final Thoughts

Maintenance excellence is not a one-time project but a continuing journey. These stories show that success comes from disciplined methods, engaged people, and leadership that values learning over quick fixes.

As you move toward the next chapter, which introduces implementation roadmaps and maturity models, consider where your organization stands on this journey—and how these examples can illuminate the path forward.

Reflection

Looking back over these projects—and many others like them—what stands out most is how similar the underlying challenges are, regardless of industry or geography. Whether it was mining, energy, manufacturing, or healthcare, the barriers to maintenance excellence were rarely technical. They were human. People's beliefs about what maintenance is, how value is created, and what "good" looks like determined success or failure far more than any technology ever did.

In every case, progress began when people saw *why* the change mattered. Sometimes it was a small pilot that produced undeniable results; sometimes it was a leadership decision to treat maintenance not as a cost, but as a capability. Once that mental shift occurred, the rest—process design, systems, data, standards—followed more naturally.

Organizational alignment also kept short-term pressures in check. Where reactive thinking dominated, improvements stalled. Where leadership insisted on long-term reliability, tactical decisions—planning, RCM, precision maintenance, analytics—could compound effectively. The cases demonstrate that reliability is not merely a technical journey, it is an organizational choice reinforced through decision-making structures.

What also became clear is that maintenance excellence cannot be imposed. The consultant's role—my role—is to create the conditions for others to learn, to see their own opportunities, and to believe they can sustain them. When people own their improvements, they don't need ongoing external support; they become self-improving organizations. That, ultimately, is the goal.

Finally, these experiences reinforced the broader truth embedded throughout *Steadfast*: Excellence in maintenance is not about perfection—it's about purpose. The best programs balance rigor with pragmatism, and they adapt to their environment rather than copy someone else's "model." Each case in this chapter reflects a journey toward that balance, where technology, process, and people converge to serve the greater purpose of asset reliability and organizational performance.

CHAPTER 18

ROADMAP TO STEADFAST PERFORMANCE

From Reaction to Strategy

Every maintenance organization is on a journey, whether it realizes it or not. Some live in a world of emergencies—reacting to whatever breaks next. Others have learned to plan and control their work, but still struggle to get ahead of the curve. A few have matured further, using proactive maintenance, data, discipline, and foresight to drive out chronic problems before they happen. The most mature are those who treat asset reliability as a key enabler of business strategy, not just a maintenance activity.

This chapter maps that journey. It's not a rigid formula but a *roadmap*—a way to see where you are, where you're headed, and what must change to reach the next level.

To make this journey easier to see, it helps to visualize the progression. Every organization moves through predictable stages as it shifts from reacting to failures, to planning work, to enabling decisions with data, and ultimately to treating reliability as a strategic capability. Figure 18-1 summarizes those stages, along with the pivotal organizational alignment that makes the shift to strategy possible.

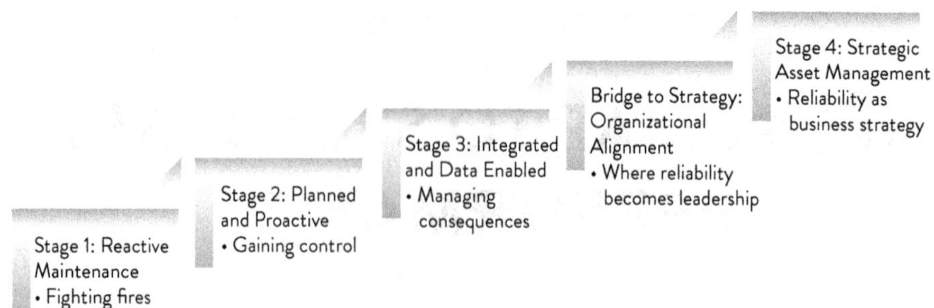

FIGURE 18-1 The Steadfast Maturity Path—the Journey from Reactive Work to Strategic Asset Management

Steadfast—Maturity Path

The path from the default—reactive stage to choosing proactive and, ultimately, strategic maintenance rarely happens all at once. Unless an organization has started up with full operational readiness, it will be somewhere between the two end stages. Each stage entails a sequence of deliberate steps, each building on the foundations of the last. The path begins wherever you already are on that journey—you can always improve. How far you go will depend on your choice of the performance required to achieve your business strategy.

- **Stage 1—Reactive Maintenance.** Maintenance responds to failures; chaos dominates and costs escalate.
- **Stage 2—Planned and Proactive Maintenance.** Work is scheduled and controlled; discipline reduces unplanned downtime.
- **Stage 3—Integrated and Data-Enabled Maintenance.** Processes, teams, and information come together to manage consequences and prevent recurrence.
- **Bridge to Strategy—Organizational Alignment.** Leadership structure, decision rights, and governance enable the pivot from operations to strategy.
- **Stage 4—Strategic Asset Management.** Reliability becomes a business capability, informing decisions from the shop floor to the boardroom.

The destination—*Steadfast* performance—isn't just about predictive technologies or "smart" assets. It's about a transformation, how an organization thinks about maintenance, reliability, and value creation, and how it uses that thinking to drive behavior on the shop floor and in the boardroom.

Stage 1: Reactive Maintenance—Fighting Fires

In a reactive environment, maintenance happens when something fails. It's all about repairs. Crews are constantly in motion, yet the same breakdowns keep recurring. The work is heroic—maintainers take pride in "saving the day"—but it is exhausting. Costs are unpredictable and high, downtime is frequent and long, and safety risks are ever-present.

The reactive environment is chaotic, high cost, and unsafe, yet it is natural. If you do nothing to actively change it, you will remain in that state, and it will only get worse.

This mode of operation often grows organically in plants that have matured faster than their maintenance systems. Equipment gets added, production pressure rises, and the maintenance team adapts as best it can. Over time, firefighting becomes normal. The irony is that many organizations in this state believe they are "too busy" to plan—when they are busy precisely because they don't plan. *Busyness* is not a sign of importance; it's often a symptom of unmanaged work. High costs and lack of budget become excuses for not investing in doing the right maintenance the right way, the very things they need to lower costs.

The first step out of this mode is stabilization. I've led large-scale transformations that began with root cause analysis on the big budget and time thieves—equipment and systems that were unreliable and burning up resources. It creates a small, stable platform from which to launch the journey.

At first, you'll know what is costing the most time and manpower and causing the greatest losses. But soon you'll need deeper insight to keep up the initial momentum. You'll need to understand which assets are most critical if they fail. What are their impacts on safety, environmental compliance, production, and other business risks?

Then you will need to understand just how well they are, or aren't, performing. What is their availability and what should it be? Is availability low because they break down too often or because they take too long to repair? Information becomes critical to answering the important questions. Do you have the right data to inform decision making?

With hundreds or thousands of individual assets to manage, you'll need a management system. A simple computerized maintenance management system (CMMS), or even a spreadsheet at first, can help in tracking failures, costs, and work history. Equally important is bringing some order to spare parts management. If you've got stashes of parts hidden away by maintainers, then you've got big problems. Knowing what you have on the shelf (and what you don't) eliminates a surprising amount of wasted time. In a late 1990s time study conducted across multiple industrial sites, we found that upward of 14 percent of a maintenance technician's day was spent simply looking for parts. It was one of the largest identifiable time drains. Subsequent industry benchmarks[1] continue to show that wrench time in most organizations remains below one-third of the available day. At this point in the journey, having too much on the shelf is cheaper than the downtime you will suffer while your trades are looking. You can optimize inventory later; first, stabilize operations.

With a bit more stability, less chaos, you can begin implementing the maintenance basics: lubrication, cleaning, and inspection routines. These may sound simple, but they form the foundation for everything that follows. Lubrication is the life blood in any machine. Cleaning prevents contamination, keeps temperatures down, and exposes small defects before they become big problems. While cleaning, people see defects when they are small (a leak), before they become big problems (a broken gearbox). When people see that small, consistent efforts prevent big crises, the culture begins to shift. They like the reduction in chaos and gain confidence that you are now applying what they will see as simple common sense.

The goal at this stage is not perfection; it's stability and control. Stop the bleeding, get visibility, prove that simple steps can have big impacts, and start managing rather than reacting.

Stage 2: Planned and Proactive Maintenance—Gaining Control

Once the fires are somewhat under control, the next challenge is to stay ahead of them. This is where proactive maintenance (PM) and planning come into play.

The organization starts scheduling regular criteria-based inspections, lubrication, and replacements of known wearing components based on time or usage. Failures become less frequent, and downtime starts to decline.

This stage introduces a new kind of discipline. Proactive work is planned in advance, and schedules are coordinated with operations. The planner's role emerges as a vital one: ensuring that technicians have the right tools, parts, and instructions before they start a job. That single change—*planning before doing*—has an enormous impact on workforce efficiency, job quality, and downtime.

Proactive maintenance includes inspections (predictive or condition monitoring), preventive (replacements and restorations), and detective (testing). Preventive maintenance is easily overdone: Equipment manufacturers recommend it because (1) they often don't know any better because they don't actually maintain what they manufacture, and (2) it encourages lucrative aftermarket parts' sales. Manufacturers often earn higher margins on parts (which you need to buy from them) than they earn on the original equipment sales (for which they had to compete), and they encourage this with warranties conditional on following their recommendations. This is why blindly following OEM recommendations can inflate workload and cost without improving reliability.

Your mechanics will see that parts they are replacing are in nearly as good condition as the new ones. They'll know that those parts could have lasted longer, often much longer. They'll perceive it is wasteful, thinking the company is doing well if it can waste so much money, but they may not question what appear to be well-informed decisions.

Inspections can easily get out of hand. If there are no clear acceptance criteria, any little flaw will be seen negatively and trigger a replacement. Many flaws are nothing more than normal wear and tear and early in the component's useful life. When a maintenance task is to "inspect and repair as required" without any clear acceptance criteria, it is a huge red flag that you are wasting effort and money, and incurring far more downtime than you really need.

Many proactive programs become overloaded with tasks like those. Many tasks add little or no value, and some can even destroy it. A motor might be torn down for inspection every year even though it's been running perfectly, while other assets are neglected. The answer isn't to abandon proactive maintenance, but to refine it. This is where the logic of reliability-centered maintenance (RCM) begins to enter the conversation. It helps you define the right maintenance for your proactive program.

Data from the CMMS helps identify which failures are most frequent or costly, and which PMs are effective. This may seem obvious, but if you spot trends in failures after you've done PMs, you should probably stop doing those PMs. Many failures are actually caused by disturbing the equipment. As one maintenance superintendent put it, "Whenever something breaks, the first place to look is the last place that someone worked."

Ask your operators about that. Many can attest to the fact that, time and again, something that maintenance just worked on isn't working as expected or needed. You've probably been through plant shutdowns or even just taken your car to a garage. Does the plant work well when it starts up or does it have problems? Have you ever had to take the car back for something they missed or a new problem that wasn't there before?

Educating your team in PM techniques and where they are most appropriate goes a long way. Over time, the team learns to optimize its program—focusing on the right work at the right time. Introducing condition-monitoring technologies such as vibration, infrared, ultrasonics, and oil analysis will help reduce intrusive inspections and target your efforts where they matter most.

At this level, leadership must reinforce a culture of discipline, precision, and following process. Set goals for planned work percentage, backlog levels, and PM completion and schedule compliance. Celebrate improvement. The message should be clear: Maintenance is not just fixing things—it's managing reliability. Don't forget root cause failure analysis (RCFA); it will still be useful at solving the breakdowns that keep happening.

As you make progress with improvements, your idea of what is "too frequent" a breakdown will change. Initially, it might be "breakdown more than three times a year." When I was a reliability engineer in a complex of petrochemical plants, that was my starting point. You will find, as I did, that in about a year of problem solving that "bad actor threshold" can be reduced to "two times" or even better.

Stage 3: Integrated and Data-Enabled Maintenance—Managing Consequences

At this stage, the organization is more or less under control. Proactive work is planned and scheduled; it is getting done on time. You are identifying problems before they reach the totally failed state. Your maintainers realize that not all failures can be prevented, but many can be found in their infancy. Downtime has been decreasing, and time between failures increasing. You still have some "bad actors" (equipment that fails more often than you can tolerate), and although operation is more stable, it has a way to go to reach benchmark performance levels. It's good, but not great.

Planning has expanded to recurring repairs. Execution efficiency is up. Schedule interruptions because of breakdowns are fewer and declining.

Your proactive program has been improved dramatically through RCM. You discovered that operating practices can be improved, previously mysterious events are now understood, and safety risks you hadn't recognized are reduced. You learned that not all failures matter equally; there is no need to treat them all as emergencies. As your operators and maintainers learned RCM and applied it working together in teams, they gained deeper understanding of each other's knowledge, and mutual respect grew. The old operations and maintenance animosity eroded.

CASE STUDY

Wastewater Utility—RCM Reveals Design Flaw

In one wastewater treatment utility, RCM revealed the cause of mysterious hazardous gas leakage events that had previously required emergency shutdown and gas-clearing efforts to avoid health problems, unwanted odors, and explosion risks. A simple piping change on an incorrectly designed drain system eliminated what once had been an alarmingly frequent event.

CASE STUDY

Cobalt Refining—RCM Reveals Missing PM and Operator Misunderstanding

In a cobalt refining process, RCM revealed the cause of a failure that often cascaded into a plant shutdown. In describing the effects of the failure, they realized that operator overcorrections amplified the issue; revised procedures made cascading failures a thing of the past. As it happened, the original cause of the failure could be easily managed with a maintenance check, so in combination the chances of that cascade ever happening again disappeared.

As stability grows, condition-based maintenance (CBM) becomes the natural next step for your critical assets. You are starting to use sensors and analytic tools to provide real-time feedback about equipment condition. Proactive repairs are now being triggered by evidence, not a calendar, and breakdowns are less frequent. This marks a profound change in thinking: Maintenance is becoming *data-enabled*. Failures are anticipated and intercepted before they occur.

But predictive capability alone does not make an organization proactive. True proactivity comes from *learning*. Every significant failure is investigated—not to assign blame, but to understand causes and prevent recurrence. RCFA becomes routine. Precision maintenance practices, such as alignment, balancing, and fastener torque control, are embedded into every job. You have lubrication technicians who manage lubricant storage, cleanliness, and their correct application. Over time, chronic problems simply disappear and energy consumption declines as described earlier and noted in the U.S. Department of Energy operations and maintenance best practices paper.

Your use of information systems should evolve as well. You may realize that the CMMS you acquired wasn't set up properly: It ran, but it didn't serve you well. You may alter its configuration or upgrade it. You are realizing that it is a tool to help in managing a process and your process needs to be refined. Your planners are using the tool, but follow-on scheduling is being done in isolation from spares availability. You need planning and work forecasting to inform spares inventory decisions so spares support can be streamlined and inventory levels brought into alignment with real demand. Nothing should be scheduled without knowing that all the parts are actually ready.

Stores won't need to be overstocked. When you add new equipment and know what parts are needed for it, you can provide inventory management with a forecast. With accurate demand based on consumption and planning forecasts, you can move away from just-in-case thinking and make truly informed decisions about stock levels, what to stock, what is actually no longer needed, and what fast-moving parts and supplies can even be managed by vendors. You'll find that fasteners, pipe fittings, electrical components, O-rings, seals, bearings, and many filters can all be moved off your shelves into vendor-managed inventory.

Managing equipment condition usually requires a combination of analytic software (e.g., vibration and oil analysis), equipment performance monitoring (production control systems and data historians), and the history of the equipment as found in your CMMS. You might have specialty software for RCM, RCFA, and FRACAS.[2] There is an ecosystem of systems used in managing your assets—some are integrated together and others are standalone.

Maintenance, operations, and engineering begin to collaborate as equal partners in creating and sustaining asset availability. The organization is no longer asking, "How do we fix it faster?" but "Why did it happen, and how do we make sure it doesn't happen again?"

Leadership at this stage focuses on integration. Encourage cross-functional teams, reward problem-solving, and make asset availability a shared KPI. As the organization matures, maintenance becomes less visible. It is still critically important, but so well managed that problems seldom reach crisis stage. You've put the energy into maintenance, and chaos has become a thing of the past.

Costs are down; productivity and production are both up. Safety has improved and business risks like environmental compliance, reputation, and continuity are well under control. Retention improves. As chaos recedes, you discover you can sustain control with fewer trades, because work is planned, predictable, and safer.

Organizational Structure and Governance—the Hidden Accelerator

Up to this point in the journey, most of the work is operational: stabilizing chaos, improving planning, refining PMs and RCFAs, adopting condition monitoring, and building discipline. But the pivot to strategic reliability cannot occur on operational momentum alone. It requires organizational structure.

In every successful transformation I have seen—as well as in the case studies in the previous chapter—the turning point came when leadership established a clear governance framework. This included who makes reliability decisions, how priorities are set, how operational pressure is balanced against long-term asset stewardship, and what forums connect operations, maintenance, engineering, and supply chain.

Organizational structure does not need to be complex. It might be a steering committee, a reliability leadership team, or a corporate standard with local ownership. What matters is that reliability has a seat in decision making and that strategy is translated into repeatable practices. Without governance, even the best technical advances can fade under short-term pressure.

As organizations mature, structure becomes the quiet enabler of performance: planners protected from disruption, reliability engineers empowered to address chronic issues, operators involved in equipment care, and executives who see reliability as a value stream rather than a cost center. Governance accelerates everything that follows and prevents reversions to chaos.

Structure is not overhead, it is infrastructure—the architecture that protects long-term value.

Stage 4: Strategic Asset Management—Reliability as Business Strategy

At the most mature level, maintenance and asset management are finally aligned with business strategy. This alignment is rarely accidental; it reflects deliberate governance, structure, and leadership commitment. Decisions about maintenance frequency, spare parts stocking, and equipment renewal are made not in isolation, but in the context of risk, cost, and performance.

This is the realm of *Steadfast*. Reliability is treated as a *value stream* in its own right. Digital tools such as advanced analytics, digital twins, and business intelligence dashboards allow leaders to model and monitor asset performance across its life cycle. Maintenance is no longer a cost to be minimized; it is an investment that safeguards profitability, safety, and resilience.

Organizations at this level often operate within, or informed by, frameworks like ISO 55000. They think in terms of life-cycle cost and risk management, not just maintenance budgets. The integration of data across CMMS, ERP, and IoT systems provides visibility from the shop floor to the boardroom. Predictive algorithms may flag potential risks well in advance, allowing maintenance, operations, and supply chain teams to plan around them.

You can feel it: the culture is different. Everyone, from executives to operators, understands their role in sustaining asset health. Improvement is continuous, driven by evidence and reinforced by leadership. Maintenance personnel evolve into reliability technicians and engineers, and planners become asset strategists.

In short, maintenance has become strategic, not reactive, not just proactive, but *integrated*. The organization makes decisions about capital investment, production scheduling, and supply chain design with reliability as a central pillar. That is the essence of *Steadfast* performance.

The Journey Continues

Moving from reactive to strategic maintenance can be a multiyear journey. Each stage demands new tools, skills, and mindsets, and it offers new rewards. Organizations that stay the course find that reliability becomes self-reinforcing: As performance improves, people take pride in it, which drives further improvement. It's a virtuous cycle.

The roadmap presented here is not a rigid checklist, but a developmental path for an existing operation. The key is to understand where you are today and deliberately work toward the next level. Look for improved performance; don't chase benchmarks. You are competing against your own past, not someone else's. Each improvement in reliability adds not just availability, but *confidence*: in your people, your processes, your technology, your assets, and your data.

Ultimately, *Steadfast* is not about technology or maintenance at all—it's about leadership. It's about shaping a culture that values foresight over reaction, learning over blame, and stewardship over consumption. When that culture takes hold, the mechanics of maintenance become almost invisible. What remains is performance: consistent, reliable, and aligned with purpose.

What About Greenfield Operations?

Apply *Steadfast* leadership during design, construction, and commissioning. Start early and you start right—no transition, no change-management drag. You start off with operations on the right foot.

Earlier in my career, I worked on a major shipbuilding project for the Navy. We were building 12 gas-turbine and diesel-driven, state-of-the-art missile-frigates to replace an old fleet of 14 steam-turbine-driven destroyers with vacuum tube electronics. It was to be a massive update in capability and availability—the new ships were to be more available for seagoing operations, so 12 could replace 14 without loss of mission time.

The support program was called "integrated logistics support" (ILS)—a very military term; today, in industrial applications, I'd call it "operational readiness" (OR).

Operational Readiness

We applied RCM to the evolving design of the ships from concept through to detailed design. Initially we identified system-level flaws that required design changes—easily done while still on the drawing board (we used paper drawings in those days). As design firmed up and equipment specifics were identified, we got more detailed in our RCM analysis and focused more on failure management for likely failure modes. For every failure there was a need for repairs—and plans. For every proactive action to deal with consequences, there was a need for a procedure. Identified human errors led to training, procedures, documentation, or practices.

Once all those were identified, we planned the work, designed the training, wrote the procedures, and defined the training requirements. Marine engineering technicians were moving from manual and pneumatic controls to electronics. Combat systems technicians were moving from vacuum tubes with lots of cooling, "complicated troubleshooting" to "solid state" with self-checking circuits and digital displays. Weapons technicians moved from loading guns to maintaining complex missile guidance systems.

Job plans identified parts, tools, test equipment, trades, knowledge, procedure, technical information, and skills that would be required. All of those could be put in place before the ships were launched. Crews walked onboard for the first time knowing exactly what to do, how to do it, and what they needed to do it. Spares were modeled across the fleet to account for work done onboard, ashore in a dockyard, and for a few "insurance" items that might be stored just a few hours flight away from either coast. Initial spares cost only 5 percent of the capital cost[3] of the fleet. The cost of all that up-front preparation was less than 3 percent of the capital cost[4] of a single ship.

"Integrated logistics support" is still used in defense projects. Throughout industry, the name is "operational readiness" and some of the decision criteria are a bit different, but no less important. Production and capacity are more important than time at sea and on station, but availability still matters.

Projects, Start-Up, and Creation of Chaos

Few projects are completed on time and on budget. Even fewer start up trouble-free. There are mistakes by operators and maintainers as they are doing their jobs in the new operation for the first time. Some of those mistakes are costly: Handover from the contractors to operations is often poorly handled. It is often likened to tossing the new operation over the fence. "Here you go, good luck." Project success is marked at completion, not the operation's performance.

There will have been training in what were obviously new technologies or operational processes and in how to run the control systems, but little else.

A good deal of engineering has been done. Chances are the project is late and overbudget, and there is a huge amount of pressure from finance to get it operating and generating revenue. By handover, many projects are not even complete—anything left to be done when you reach that arbitrary and invariably rushed handover date is now left to maintenance to finish.

Completions and correction to defects will be done as the operators are getting ready to start it up. You have chaos with a veneer of management control. As you start it up, a lot will go wrong, and some of that will result in breakdowns and the need for repairs. What looks like a new operation will go from quiet to chaos quickly. A lot of lessons will be learned; the learning curve is steep. Some lessons will be remembered, some documented, and many forgotten. The word-of-mouth training begins as operators have trouble, figure it out, and pass on their learning. Sometimes they pass on what will later become bad habits. Operators will gradually become proficient in keeping it running, but may not truly know how the plant should be operated.

I was involved in commissioning of a linear low-density polyethylene plant. Operators got used to minor upsets and would leave product in the process reactor vessel rather than clear it out. They could start up again, quickly and without difficulty. However, we had a major upset and they did the same thing. They couldn't start up and learned the hard way that it didn't take long for the product to solidify. It took three days with chainsaws to clean it out.

If your maintainers got involved before handover, you might know what "commissioning spares" are available and even where, but those maintainers will be well aware that they don't have enough of the right parts for what's about to happen. Those parts may or may not be absorbed into your spares inventory. A lot of rush buying will occur, much of it driven by breakdowns and demand for repairs. The maintainers, seeing how unprepared things are, will buy and stash extras away for the next time they are needed. Any hope of integration of planning and spares management is trashed.

That chaos will continue for months if not a year or longer. It takes very little time to get into a chaotic state—the starting point in the stages described in this chapter. It can take years to get out of it. It can also be avoided by that little investment in operational readiness.

The chaos arises because we manage in silos and finance projects not performance. We cut costs in projects to stay within budgets and achieve desired return on investment. But we don't invest in sustaining the very capability we are paying to build. By not funding operational readiness, we are transferring cost from capital to operating expense, and it will be much higher than it needs to be for a long time if the operation ever fully recovers.

Engineering is often outsourced to a contractor who has no stake in performance of the asset or sustaining its performance over the many years it will be in operation. You may have outsourced engineering, but you've in effect abdicated your responsibility when it comes to performance and sustaining it.

That is not managing assets. There is no life-cycle view to balancing costs, risks, and performance when we do that.

Getting to *Steadfast* Performance

While this roadmap offers clear direction, achieving *Steadfast* performance in practice requires more than process knowledge. Success comes faster when guided by professionals who have walked this path before: people who can translate strategy for the trades, and operations for executives. Partnering with experienced advisors can help organizations avoid common pitfalls and sustain progress.

There are two paths you can follow for your new assets: operational readiness or you can plunge into chaos and claw your way back from it over the next few years. This chapter describes the latter in detail because it is what most operations must go through.

If you invest in operational readiness, you may still have challenges as you commission and go into operation, but they'll be minor in comparison with what you might have otherwise. You'll be starting up prepared and making continual improvements as described in Stage 3 earlier. Operation will come up to capacity quickly and remain stable. Revenue will achieve forecast sooner.

You can save that initial investment, but you will be disappointed in results and start your *Steadfast* journey at Stage 1, after a rapid descent into chaos in your new operation.

That descent occurs in large part because of the lack of operational readiness from the start. As described earlier, the journey from handover to operational stability can be rocky. Your people will be rushed to generate revenue despite the setbacks. They will do their best in the circumstances, get very creative, and eventually get it all smoothed out, likely after a long period. They will quickly forget that those circumstances (and all their attendant consequences) could have been avoided, but they might remember that project was a nightmare.

Older operations have usually been through the initial nightmare, dealt with a lot of the problems, but even years later may still be in chaos. They've gone beyond the time when they can blame the contractors or the project for lack of preparedness. They've normalized the chaos and very likely resigned themselves to it. The signs of that are obvious in the way people talk. "Oh, that's just how it is here."

Operational readiness gets you into Stage 3 of that journey to *Steadfast* performance right from the start. Avoiding it and saving that initial investment will get you to the start of Stage 1 quickly. The investment to get out of that state will be huge. The costs include:

- All the analysis, planning, and purchasing of spares
- Training and everything else identified in operational readiness
- The production and revenue losses of a slower start-up and ramp-up to full capacity (assuming you can reach it at all)
- The losses due to more frequent breakdowns because you don't start with a good proactive program

- The long duration repairs because you don't have the planning and spares in place

What Is Operational Readiness Worth?

Each industry is quite different, and various factors will influence maintenance costs and the value of lost production.

Here are some examples of the cost to maintain heavy industrial operations and what that lack of operational readiness can cost. When looking at maintenance costs, consider that various factors can influence the numbers: age of the asset, operating environment, age of the technology/obsolescence, maintenance strategy, what they consider to be maintenance (sometimes capital improvements are included), how replacement asset value (RAV) is defined, and whether they include operating losses (labor and materials) in the costs. The RAV numbers shown are ranges that could be identified through various sources. They are best used in context with other metrics described in Chapter 7.

Lower maintenance costs are a result of the use of *Steadfast* concepts, including integrated processes, precision maintenance, and proactive maintenance defined using RCM. With operational readiness, you can get to the lower end of the range and do it more quickly. Ongoing costs will be lowered and the added productive capacity substantial.

The start-up times in the cases described below reflect getting the operations up "mechanically"—that is, running and not necessarily producing quality product at full capacity. That often takes longer. To reach high capacity with quality production requires the machinery and systems to be worked efficiently producing to specification. That can only be achieved with good maintenance practices.

Oil Refining, Petrochemicals Continuous Process Plants

Maintenance costs will be as low as 2 percent of replacement asset value but as high as 5 percent if poorly maintained. Without operational readiness, you'll be at that 5 percent level for sure, and likely for a few years, until you climb from Stage 1 chaos to stability in Stage 3. That additional 3 percent per year will total 9 percent of RAV spent instead of the 3 percent or so up front on operational readiness.

Add to that the cost of production that is lost due to all the breakdowns and downtime, and you'll see how operational readiness can easily pay for itself.

The Cabinda[5] refinery in Angola started up in late 2024 and took over six months to reach capacity. Marathon Martinez[6] (renewable diesel production) took nearly 12 months to get from 50 percent to 100 percent capacity in 2024.

McKinsey[7] reports that unplanned outages reduce capacity, increase costs, and lead to lost profit opportunities of $20 million to $50 million per year for midsized refineries.

Pulp and Paper: Mixed Batch and Continuous Process

Maintenance costs in pulp and paper range from 3 to 8 percent of RAV (7 percent to 15 percent if you consider just the production equipment). Mill start-ups are, like refineries, measured in months, not days.

Baker Huges (Bentley Nevada)[8] estimates that unplanned downtime (breakdowns) will cost $10,000 to $50,000 per hour, with a four-day outage resulting in 10,000 tons of lost output. They cite a technical brief that claims 60 to 80 percent of malfunctions are attributable to incorrect or insufficient maintenance. European and US pulp was priced between $1,600 and $1,800 per ton in 2025. At those prices, that four-day outage costs between $16 million and $18 million.

Automotive: Discrete Manufacturing

These plants have a greater diversity of equipment types than processing operations and often go through changeovers when switching from one manufactured part to another. This leads to higher maintenance costs than well-operated processing plants in the range of 3 to 6 percent of RAV. Maintenance costs are often in the range of 10 to 15 percent of operating costs.

Allaboutlean[9] indicates that automotive plants typically get up to speed initially in about six months or more. Downtime due to poor maintenance averages around 29 hours per month at high (unspecified) costs per hour. Anywhere from 1 to 10 percent of available production time can be lost when reliability is poor.

Mining: Combination of Mine and Ore Processing

Mining operations can be complex. They can be above ground, underground, or both, and may or may not include related ore processing and refining process plants. Terry McNulty produced a set of four curves that provide a well-known benchmark for mining start-ups, citing some 41 case studies. Operations deemed successful take a few months, while those that are less successful can take up to several years. Those curves were updated[10] and presented to the Conference of Metallurgists in late 2024. For ore-processing plants, it is not uncommon to take years to get to 90 percent of capacity.

The large copper concentrator at Quellaveco, Peru, started commercial operation in late 2022 and in continued ramp-up is described in public updates[11] into 2025.

Most mines target availabilities of 85 percent for their mobile equipment fleets. Poor maintenance practices can easily drag that down by 5 to 15 percent, representing a 5 to 15 percent loss of output. A rope shovel can cost roughly $60,000 per hour while down, a 200-ton haul truck around $3,000 per hour. While the shovel is down, the haul trucks that it would have been filling and their operators are idled.

Maintenance costs in mines are less benchmarked, but they are very likely comparable to, or worse than, those for aircraft[12] at around 12 percent of RAV annually. Those costs are typically in a range of 30 to 50 percent of operating costs, which is quite high but understandable given the rugged operating conditions, often in remote locations with higher support logistics costs.

AFTERWORD

FINISHING THE ROADMAP, BEGINNING THE WORK

Every successful transformation I've seen has had one thing in common: guidance from people who've done it before. The best partners bring credibility with technicians and fluency with executives, ensuring the lessons of *Steadfast* become living practice rather than just an aspirational framework.

Steadfast is not a destination you arrive at and frame on the wall; it's a way of leading. Across these chapters, we've moved from the noisy urgency of reactive work to the steadier cadence of planned and proactive practice, then onward to integrated, strategic reliability. Along the way, the tools changed—CMMS, organizational structure, condition monitoring, AI and analytics, work planning, spare parts logic—but the real shift wasn't technological. It was managerial. It was cultural.

If there's a single theme, it's this: Reliability is a choice. Not a guarantee, not a gift, not a lucky break—it's an organizational choice made and remade every day. You choose it when you stabilize chaos instead of celebrating heroics. You choose it when you define "good" clearly, and then let evidence from your data—not habit—tell you what to do next. You choose it when your planners protect the integrity of the schedule, when your technicians practice precision, when your operators own the health of the assets they run. And you choose it most of all when leaders insist that safety, quality, and asset performance aren't trade-offs, but mutually reinforcing outcomes of thoughtful design and disciplined execution.

This roadmap has also been about respect: for physics, for process, and for people. Precision maintenance respects the physics: alignment, balance, torque, cleanliness. Nature always gets the last vote. Good processes respect time: planning before scheduling, before doing; investigating before blaming; standardizing before optimizing, because time is the one resource you never get back. And culture respects people: the craft of the trades, the judgment of operators, the knowledge of the engineers, the responsibility of leaders, because reliability is built by humans who take pride in doing things right.

There is, of course, no single path. Your starting point will be different from others'. Some will inherit greenfield opportunities; most will inherit brownfield realities. Some will have strong data and weak discipline; others the reverse. What matters is not the specific sequence of steps so much as the momentum you build by taking the next right step: stabilize what's unstable, make the invisible visible, remove the chronic causes of pain, and connect decisions to risk and value across the lives of the assets.

If you remember nothing else, remember the flywheel: small, consistent improvements—all done with precision, learned rigorously, and spread through coaching—will compound. A year from now, the "bad actors" list will be shorter. Two years from now, the argument about who "owns" reliability will feel quaint. Three years from now, capital decisions will reflect life-cycle thinking, and your best people are will be teaching the next generation not just how to fix, but how to think.

The *Steadfast* journey is like a flywheel: a self-reinforcing cycle where purpose, governance, disciplined work, data-enabled learning, performance, and culture amplify one another as shown in Figure A-1.

Call to Action

So, what is your next move, right now, at the end of this book? Name the gap that hurts the most. Put numbers to it. Stand up the smallest possible cross-functional team with the authority to change one process and the discipline to measure the result. Apply root cause analysis; find the problem, fix it. Close the loop with process and performance metrics that show measurable results. Bake the learning into standards. Then do it again. Not 50 projects—one. Then another. Then another. The journey to *Steadfast* performance is simply the habit of making tomorrow's failures optional.

Grow those projects into a whole program.

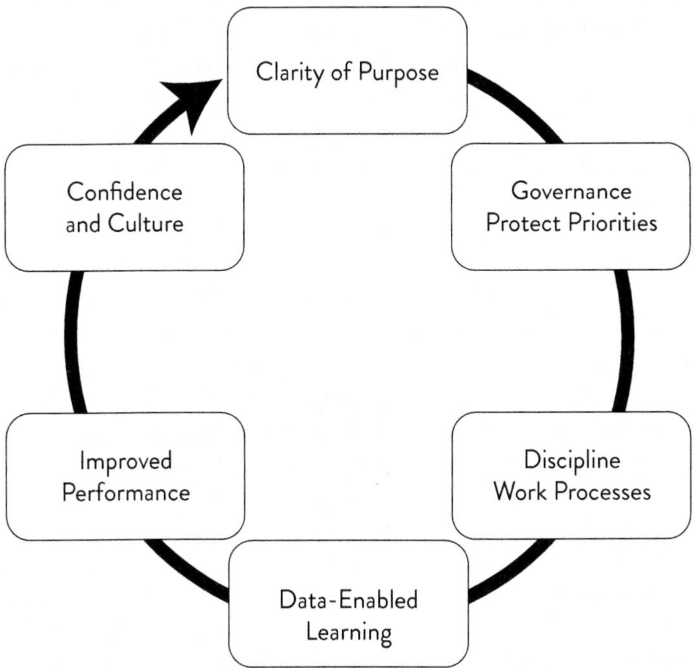

FIGURE A-1 The *Steadfast* Flywheel—a Self-Reinforcing Cycle of Reliability Excellence

As you close this final chapter, consider the quiet markers of progress: fewer emergencies, calmer mornings, jobs done right the first time, operators who notice early, planners who are believed, stores that are trusted, dashboards that inform rather than entertain and deceive. Consider the pride that grows when people see that their craft matters and their decisions are both heard and they compound. Consider the freedom that comes when performance is predictable, risk is understood, and strategy is served by the assets rather than constrained by them.

That is *Steadfast*: not a program, but a posture. Not a moment, but a movement. Not the end of a book, but the beginning of better work.

LIST OF ACRONYMS

ACA: asset criticality analysis

AI: artificial intelligence

BOM: bill of material

CBM: condition-based maintenance

CEO: chief executive officer

CFO: chief financial officer

CMMS: computerized maintenance management system (software)

CNC: computerized numeric control

COO: chief operating officer

DM: detective maintenance (testing to look for failures)

DNA: deoxyribonucleic acid (biology)

DOE: Department of Energy (US government department)

EAM: enterprise asset management (software)

EBITDA: earnings before income taxes, depreciation and amortization (financial term)

ERM: enterprise risk management

ERP: enterprise resource planning (software)

ESG: environmental, social and governance

FRACAS: failure reporting and corrective action system

GenAI: Generative Artificial Intelligence

GM: general manager

HR: human resources (department)

IIoT (IoT): industrial internet of things

ILS: integrated logistics support

IMS: integrating management system (not software)

IT: information technology (department)

ISO: international standards organization

KPI: key performance indicator

LCC: life cycle cost

LED: light emitting diode (electronic component)

MARC: maintenance and repair contract

MDT: mean down time

MRO: maintenance repair and overhaul

MTBF: mean time between failures

MTBM: mean time between maintenance (activities)

MTTR: mean time to repair

OEE: overall equipment effectiveness

OEM: original equipment manufacturer

OR: operational readiness

O&M: operations and maintenance

PM: proactive maintenance (often refers to proactive work orders generated automatically by management software—CMMS, EAM, ERP)

PMO: proactive maintenance optimization (method)

PdM: predictive maintenance (includes CBM)

PPE: personal protective equipment

PvM: preventive maintenance

QR: quick response (code)

RAV: replacement asset value

RCA: root cause analysis (method, often used interchangeably with RCFA)

RCFA: root cause failure analysis (method, often used interchangeably with RCA)

RCM: reliability-centered maintenance (method)

RFID: radio frequency identification

ROA: return on asset

ROI: return on investment

SMRP: Society of Maintenance and Reliability Professionals

SOP: standard operating procedure (document)

TPM: total productive maintenance

TPS: Toyota production system

UK: United Kingdom (country)

US: United States of America (country)

USD: US dollar

VFD: variable frequency drive

VMI: vendor-managed inventory

NOTES

Chapter 1

1. Stan Nowlan and Howard Heap, "Reliability Centered Maintenance," Report A066-578, United Airlines for the Office of Assistant Secretary of Defense (Washington, Dec. 29, 1978).
2. John Moubray, "RCM II Reliability-centred Maintenance" (Oxford, UK: Butterworth-Heinemann Ltd., 1991).
3. Anthony M. Smith, "Reliability-Centered Maintenance" (New York: McGraw Hill, 1993).
4. This is an important aspect in lean manufacturing, and several business authors and teachers emphasize its importance in achieving better results from measured approaches.

Chapter 3

1. John D. Campbell and James V. Reyes-Picknell, *Uptime: Strategies for Excellence in Maintenance Management*, 3rd edition (New York: Productivity Press, 2015).

Chapter 4

1. John D. Campbell and James Reye-Picknell, *Uptime: Strategies for Excellence in Maintenance Management*, third edition (New York: Productivity Press, 2015).
2. CMMS will be used to refer to the various systems that are out there, best-of-breed CMMS, EAM, ERP, or other integrated packages, whether they are deployed over the cloud or on your own networks.
3. This is a result of Parkinson's law: the principal that work will expand to fill the time available for its completion.
4. Terminology: Planning is about what, how, and who; scheduling is about when. Many maintainers use the term "plan" when they are referring to a schedule. For clarity and consistency, in this book care is taken not to mix the two up.
5. Criticality is a measure of how important the equipment is to the operation based on criteria such as the equipment's impact on safety, environmental compliance, and potential revenue loss, if the equipment were to fail.

6. "The Productivity J-Curve" (NBER working paper, 2018; rev. 2020); Andrew McAfee and Erik Brynjolfsson, "Investing in the IT That Makes a Competitive Difference," *Harvard Business Review*, 2008; Erik Brynjolfsson and Lorin M. Hitt, "Beyond Computation," *Journal of Economic Perspectives*, 2000; Erik Brynjolfsson's Stanford page (bio); and Wired for Innovation slides/book also emphasize organizational capital as the driver of realized productivity from IT.
7. Wireman's CMMS and TPM books are widely cited in maintenance/reliability circles; they emphasize that process definition and redesign and change management are pre-requisites to CMMS value. His Reliabilityweb.com profile (plus third-party summaries) establishes his authority on process-first CMMS implementations.

Chapter 5

1. Stan Nowlan and Howard Heap, "Reliability Centered Maintenance," Report A066-578, United Airlines for the Office of Assistant Secretary of Defense (Washington, Dec. 29, 1978), and others.
2. John D. Campbell and James Reye-Picknell, *Uptime: Strategies for Excellence in Maintenance Management*, third edition (New York: Productivity Press, 2015).
3. Jesus Sifonte and James Reyes-Picknell, *Reliability-Centered Maintenance--Reengineered* (New York: Productivity Press, 2017).

Chapter 6

1. John D. Campbell and James V. Reyes-Picknell, *Uptime: Strategies for Excellence in Maintenance Management*, 3rd edition (New York: Productivity Press, 2015).

Chapter 7

1. Terry Wireman, *Developing Performance Indicators for Managing Maintenance* (New York: Industrial Press, 1998).
2. Diagram is from *Paying Your Way—Improving Performance Through Uptime*, by James Reyes-Picknell, Conscious Group Inc., 2020.
3. VFD is variable frequency drive used to alter the speed of drive motors.
4. SMRP publishes some nine different papers on KPIs. Ricky Smith and David Martin produced "10 Maintenance and Reliability KPIs You Need to Know and Use" in 2012.
5. James B. Humphries, VP Manufacturing Technology and Operations Support, Fluor Daniel, Greenville, South Carolina, "Best-in-Class Maintenance Benchmarks," *Iron and Steel Engineer*, October 1998. Magazine is no longer published.

Chapter 8

1. The use of job-planning templates and even AI in planning jobs are identified in Chapter 4.
2. 5S refers to the lean manufacturing practices of: sort (things are where they below), set in order (easy to find), shine (clean and ready for use), standardize (one variation, not many), and sustain (put in the effort to keep it that way).

3. BOMs: bills of materials are lists of the parts and materials in a piece of equipment or system.

Chapter 9

1. Operational and maintenance readiness programs are key to achieving this. Avoiding them and assuming that manufacturer recommendations will be sufficient is a big strategic mistake.
2. John D. Campbell and James V. Reyes-Picknell, *Uptime: Strategies for Excellence in Maintenance Management*, third edition (New York: Productivity Press, 2015).
3. Risk is a combination of consequence and probability. "Risk adjusted cost" is therefore the cost associated with the consequence you want to avoid, multiplied by the probability of it occurring.
4. PM: proactive (or preventive) maintenance.
5. EBITDA: earnings before income tax, depreciation, and amortization.
6. ROI: return on investment.
7. Accountants won't be able to verify this as the money isn't "lost," but financial managers will understand the concept.

Chapter 10

1. Ron Moore, *Making Common Sense, Common Practice: Models for Manufacturing Excellence*, third edition (Burlington, MA: Elsevier Butterworth-Heinemann, 2004), Figure 9–14.

Chapter 12

1. ISO 55000:2024 (Vocabulary, Overview, Principles), 55001:2024 (Asset Management Systems—Requirements), 55002:2018 (Guidelines for application of 55001), 55010:2024 (Guidance on the alignment of financial and non-financial functions in asset management), 55011:2024 (Guidance for the development of public policy to enable asset management), 55012:2024 (Guidance on people involvement and competence), 55013:2024 (Guidance on the management of data assets).
2. John D. Campbell and James V. Reyes-Picknell, *Uptime: Strategies for Excellence in Maintenance Management*, 3rd edition (New York: Productivity Press, 2015).
3. Canada's province of Ontario is a good example. It has its "Infrastructure for Jobs and Prosperity Act, 2015," S.O. 2015, c, 15, and regulation O. Reg. 588/17 "Asset Management Planning for Municipal Infrastructure."
4. The U.S. Department of Energy's *Operations & Maintenance Best Practices Guide* (PNNL-14788).
5. An integrated management system (IMS) integrates various management systems and processes within an organization into one complete framework. It enables operation as a single unit with unified objectives, rather than managing each in isolation, with the inherent risks of conflicts and inefficiencies.

6. "Middle Mountain Utilities" is a fictional name used to protect the identity of the actual organization being described.

Chapter 13

1. A few examples: Safety devices had been disabled at Chernobyl, various control indicators and warnings didn't work at BP's Texas City plant in March of 2005, parking brakes didn't work on the train leading to the Lac Megantic, Quebec, disaster in July 2013, and the Deepwater Horizon disaster in April 2010 occurred in part because of faulty safety devices.
2. In 2024, global data center energy use was estimated to be 415–460 TWh (about 1.5% of the world's energy consumption) and AI was recognized widely as the biggest driver of growth. In 2025 AI alone was estimated by Schneider Electric to consume roughly 15 TWh, and that grows to 100 TWh when considering educational and training uses. Forecasts by 2030 are that global data center electric demand will be on the order of 1,000+ TWh and that the broad use of AI will account for some 500 TWh.
3. DOE/FEMP, "Operations & Maintenance (O&M) Best Practices Guide," Release 3.0, August 2010.

Chapter 15

1. Governance Reminder—ISO 55000 Principle Responsibility for asset performance can be shared, but accountability cannot. Leadership must ensure that outsourced reliability arrangements remain consistent with corporate asset management policy and risk governance frameworks.

Chapter 16

1. Specifically, the second law of thermodynamics.

Chapter 17

1. "Improving equipment uptime by 40% with predictive maintenance solutions," Oxmaint, September 1, 2025, https://oxmaint.com/case-study/post/predictive-maintenance-up-time.
2. "Precision maintenance: What it is, why you need it, and how to start," UptimeAI, https://www.uptimeai.com/resources/precision-maintenance/.
3. Arnulf Hagen and Trond Michael Andersen, 2024, "Asset Management, Condition Monitoring and Digital Twins: Damage Detection and Virtual Inspection on a Reinforced Concrete Bridge." *Structure and Infrastructure Engineering* 20 (7–8): 1242–73. doi:10.1080/15732479.2024.2311911.

Chapter 18

1. Keith Mobley and Ricky Smith, "Facts about maintenance wrench time," Reliable Plant, August 2023, https://www.reliableplant.com/Read/32402/facts-about-maintenance-wrench-time.
2. FRACAS is "failure reporting and corrective action system." It is used as a continuous improvement tool to track failures and whatever follow-up actions that are taken to avoid having the failure occur as often or again in the future.
3. We identified some $600 million in spares for initial support of the $12 billion fleet (1980s).
4. The ILS part of that program was roughly $30 million, compared with the $1 billion price of one ship.
5. Reuters, "Angola's new Cabinda refinery to start up later this year – CEO," July 25, 2024, https://www.reuters.com/business/energy/angolas-new-cabinda-refinery-start-up-later-this-year-ceo-2024-07-25/.
6. "Marathon expecting Martinez RD Facility at full capacity by year end – OPIS," *Wall Street Journal,* April 30, 2024.
7. Matt Gentzel, Diego Giraldez, Bill McDonnell, and Anantharaman Shankar, "The case for doubling down on refinery reliability now," McKinsey & Company, June 10, 2024, https://www.mckinsey.com/industries/energy-and-materials/our-insights/blog/the-case-for-doubling-down-on-refinery-reliability-now.
8. "Maintenance: The hidden profit generator in pulp and paper manufacturing," Bentley Nevada, 2020, https://dam.bakerhughes.com/m/6d352ed07382121b/original/BH_BentlyNevada_PulpPaper-Awareness.pdf.
9. Christoph Roser, "Preparation for Ramp Up and Down of Production," September 5, 2023, AllAboutLean.com, https://www.allaboutlean.com/preparation-ramp-up-down-production/.
10. Terry McNulty and Krishna Parameswaran, "McNulty Ramp-Up Curves: An Update," Metallurgy and Materials Society of CIM (eds), *Proceedings of the 63rd Conference of Metallurgists*, COM 2024, Springer, Cham., November 19, 2024, pp. 917–923, https://doi.org/10.1007/978-3-031-67398-6_153.
11. Anglo-American, investor guidance prospectus.
12. Ron Moore, "Maintenance costs as a percentage of replacement asset value: A useful measure?" Association of Asset Management Professionals, June 21, 2024.

INDEX

References to figures are in italics.